ISBN 978-1-332-13523-3
PIBN 10289450

BRITISH ASSOCIATION

CARDIFF MEETING, 1920.

Handbook to Cardiff

AND THE NEIGHBOURHOOD

(WITH MAP).

PREPARED BY VARIOUS AUTHORS FOR THE PUBLICATION SUB-COMMITTEE, AND EDITED BY

HOWARD M. HALLETT, F.E.S.

CARDIFF.
MCMXX.

PREFACE.

This Handbook has been prepared under the direction of the Publications Sub-Committee, and edited by Mr. H. M. Hallett. They desire me as Chairman to place on record their thanks to the various authors who have supplied articles. It is a matter for regret that the state of Mr. Ward's health did not permit him to prepare an account of the Roman antiquities.

D. R. Paterson.

Cardiff,
August, 1920.

CONTENTS.

Cardiff and Neighbourhood.

PREHISTORIC REMAINS.

By JOHN WARD, M.A., F.S.A.

For the purposes of this sketch, the "Cardiff District" is all the land on the north side of the Bristol Channel which lies within a circle having Cardiff for its centre and a radius of 25 miles. This circle contains the eastern two-thirds of Glamorgan; all Monmouthshire, less about a fifth on the north and north-east; an irregular strip of south Breconshire, and a small fraction of Gloucestershire in the extreme east. Besides Cardiff, the region contains the important towns of Newport and Merthyr Tydfil, and the smaller towns of Bridgend, Cowbridge, Barry, Penarth, Pontypridd, Dowlais, Aberdare, and Caerphilly, in Glamorgan; and Pontypool, Usk, Tredegar, Abertillery, Blaenafon, Abersychan, Abergavenny, and Chepstow, in Monmouthshire. Some of these are towns of ancient lineage, and some are of phenomenally recent growth with names familiar in their connection with coal.

In studying the ancient remains (Roman and mediaeval as well as the earlier), a prominent feature in the physiography of the district should be kept in mind, as it has to a considerable degree determined their distribution. The Great Western main line from Cardiff to Bridgend and beyond, passes through an undulating fertile country, dotted with ancient villages and exhibiting all the signs of having long been under cultivation. It is known as the Vale of Glamorgan, and is bounded on the south by the sea and on the north by a range of lofty hills. These hills are the scarp of a more elevated country which is sculptured in bolder relief and furrowed by deep valleys with intervening heights, bare and wind-swept. These uplands stretch northwards into Breconshire with increasing elevations. Along the southern fringe, heights ranging from 800 to 1,000 feet are attained, but these are doubled before the northern border of our district is reached. Eastwards, the uplands extend into Monmouthshire to the vicinities of Pontypool and Abergavenny, and their extent is indicated by a broken line on our first map. This line is the 400-feet contour which admirably serves the purpose, as few hills in the lowlands exceed that height, and only the deeper valleys in the uplands have their bottoms below that level. The rest of Monmouthshrie which comes within our purview, has not the bold relief of Glamorgan, but there is a considerable tract between Wentwood and the Wye which rises above the 400-feet contour, but it nowhere exceeds 932 feet.

In its prehistoric remains, our district is decidedly rich. It would probably be difficult to find elsewhere in our island so many tumuli crowded within a space of the size of the portion of Glamorgan which comes within our range, or camps so thickly strewn as those of south and south-east Monmouthshire. Yet, unfortunately, it cannot be said that they have received the comprehensive and systematic study they deserve, and very few of them have been subjected to the most important implement of archaeological research—the spade. In these respects they contrast with the Roman remains which also are numerous and important, but have been extensively explored and studied, and have materially widened our knowledge of Roman Britain.

Palaeolithic Remains.—Nothing has been found in our district which has been satisfactorily assigned to palaeolithic man, although just beyond its eastward fringe, the Doward caves, and similarly beyond the western, those of Gower, have yielded important remains of that era.

Chambered Tumuli.—There are within our district seven undoubted chambered tumuli. The largest is at Tinkinswood, near St. Nicholas. The chamber is a fine example of megalithic construction, and the capstone (22ft. 4in. by 15ft. 2in.) is considered to be the largest in Great Britain, but is probably not the heaviest in consequence of its relative thinness. The mound, which has an approximately east and west orientation with the chamber near the former end, has been much reduced, especially on the south side, by the removal of stones centuries ago, and probablity was during this process that the south side of the chamber was destroyed and the contents scattered. The remains were excavated in 1914, with results that exceeded expectations. The cairn was found to be in length, 130ft., and in width, 58ft. 8in. at the eastern end and 55ft. at the opposite end, and to be enclosed in the remains of a retaining-wall faced with quarried stones neatly fitted together. On the east, this revetment, instead of returning straight as on the other sides, curved inwards at a short distance from each corner and flanked a spacious approach to the chamber, the entrance to which was near the north corner. The great stones of its front had been clothed with a sloping wall, which apparently reached the capstone, the projecting edge of which would form a rude cornice. The entrance had been closed with a roughly squared flagstone which was found fallen inwards and broken. The scattered contents of the chamber consisted of human bones, all in a fragmentary condition, with a small number of sherds of handmade pottery and several bits of flint. The bones were submitted to Dr. Arthur Keith, F.R.S., who reported that they represented at least fifty individuals, male and female, from infants to adults of seventy years of age. As the capstone was fractured and in a dangerous condition, a stone pier was introduced to ensure its safety ; and the retaining walls of the front of the tumulus were partially restored, the restoration being in herring-bone work to distinguish it from the ancient masonry. (" Arch. Camb." 1915, p. 253 ; 1916, p. 239.)

A mile to the south-east, and near Maes-y-felin farm, is a similar chambered tumulus, with a similar orientation. Although on a smaller scale, the chamber, instead of being half-buried like that of Tinkinswood, stands more revealed, and for this reason is better known than its neighbour.

West of Bridgend and between Laleston and Tythegston is a little-known tumulus of the kind. It is, in its present condition, oval, but somewhat irregular, about 80ft. in length and 6 or 7 ft. high. On the summit lies a large thick slab (about 15ft. by 6ft. 8in.), which almost certainly is the capstone of a buried chamber. It is curious that so promising a site should have received so little attention from the archaeologist and the topographer.

On the south side of the lane to Cae-yr-Arfan Farm, near Creigiau Station, are the remains of a chamber partly incorporated in the lane wall. The capstone (about 10ft. by 5ft., but originally wider) is in position, but the structure is in a ruinous condition. There is now no sign of a mound, though fifty years ago there were visible traces of one on the south. " Arch. Camb." 1875, p. 181 ; 1913, p. 111 ; 1915, p. 420.)

Near Dyffryn, on the west side of Tredegar Park, are the remains of another. The mound, which is much reduced, has an east and west orientation, and the chamber, now a mere wreck, is near the former or wider end. Three of the uprights remain standing, but the capstone has been pushed off and is partially buried in the ground. These remains, although very conspicuous, are not marked on the Ordnance Survey. (Bagnall-Oakeley, " Rude Stone Monuments, etc." p. 11 ; " Arch. Camb." 1909, p. 271.)

Near Newchurch, Caerwent, is a megalithic chamber in a semi-wrecked condition, known as " Gaer lwyd." No plan or measured drawing of it has been published and the published information is meagre. To one writer it seemed to have been surrounded with a slight ditch and bank, while according to another the base of the mound is visible on the north-west. (" Arch. Camb." I. p. 277 ; 1909, pp. 266, 271.)

On Heston Brake, a small knoll near Portskewett Station, is an interesting and unusual chambered tumulus. Here again the mound has an east-and-west orientation and is much reduced. It contains two half-buried chambers, which lack their capstones and some of their uprights. They are roughly oblong in shape, the eastern 13ft., and the western 9ft. in length, and are placed end to end and communicate with one another. As they are not truly in line with one another and are on different levels, it is probable that the western chamber was an addition to the original construction. The entrance is on the east, and flanked with two pillar-stones, one broken. Similar pairs of pillar-stones occupied the same positions in chambered tumuli at Five Wells, near Buxton and the Bride Stones, near Congleton,* but they are of very rare occurrence. The Heston Brake tumulus was excavated in 1889, when a few human bones were found in the western chamber, but the exploration was not thorough. (Bagnall-Oakeley, "Rude Stone Monuments," etc., p. 18.).

Besides the above, there are a few stones or piles of stones which have been regarded as the remains of megalithic chambers. In the immediate vicinity of the St. Nicholas tumulus are two such piles which certainly look like fallen chambers at first sight. They consist of slabs of the "cromlech bed," but as they are on the outcrops of that rock, they *may* be nothing more than exposed and weathered portions of it. A mile west of St. Nicholas is another huge slab of this rock marked "Standing Stone" on the Survey. It is in a water-worn hollow—an unlikely spot for a tumulus, —and it is more reasonable to think that it is a naturally dislodged piece of the rock, than a portion of a dolmen. (Malkin, "South Wales," I., p. 74). More puzzling is a pile of weather-worn stones, marked "Cromlech," north of Coity Castle, near Bridgend. It is at the foot of a steep slope (again an unlikely spot for a tumulus), from which the stones may have descended by gravitation; but if so, it is curious that the result should so resemble a collapsed chamber!

In the grounds of Druidstone, a modern mansion near Michaelston-Fedwy, is an upright slab, more than 9ft. high and 7ft. wide (marked "Standing Stone" on the Ordnance Survey), which has strong claims to be regarded as the relic of a demolished chambered tumulus. Although

* See "Reliquary," 1901, p. 235; and Sainter's "Scientific Rambles round Macclesfield."

there is no visible sign of a mound, the situation—a brow—is favourable. But more to the point is the uncouth name, Gwal-y-filast (kennel of the greyhound bitch), of the old farmhouse near by. Now this is a common name applied to dolmens in South Wales, and has an early origin—(see "Arch. Camb." 1876, p. 236, and 1915, p. 282.) According to Lewis's "Topographical Dictionary" there were formerly dolmens at Marcross and Llangynwyd, in each case known as "The Old Church."

Other Tumuli.—As the first map well indicates, the chambered form a very small proportion of the total of ancient burial mounds in our district. This map is compiled from the Ordnance Survey, mainly from the 6 inch-to-1-mile sheets, as they show more of these remains than the smaller scale issues, but a few of them are inserted from personal observation and other sources of information. For some inexplicable reason, these tumuli have received but scant attention from local antiquaries, for the number which have been excavated in the interests of science is less than a dozen, and to these can be added several accidentally discovered interments (no doubt originally covered with mounds) which have been placed upon record. As these described examples belong to the Bronze period, it is reasonable to conclude that the majority, at least, of the tumuli we are considering are also of that period. So far as I gather from the 6in. Survey and from my own observations, they are, with few exceptions, of the ordinary circular bowl-shaped form, but some have been pulled about, probably by old-time treasure-seekers. They differ greatly in size, but the smaller predominate. Most appear to be cairns, but the chinks between the stones are usually filled with earth, probably derived from an earth-cap or blown in as dust. Cists appear to have been usual, and they may be seen exposed in some of the mutilated cairns.

Broadly speaking, the tumuli are most thickly strewn in the hill county of Glamorgan, whereas in the Vale (with the exception of the district between Cowbridge and Nash Point) and in the parts of Monmouthshire below the 400ft.

contour, they are few. Again, they are confined, with few exceptions, to summits and brows. This distribution is due in considerable degree, to the unequal advance and intensity of human settlement. The Ordnance Survey shows that the tumuli are by no means evenly distributed on the heights. One hill may be sprinkled with them, while its neighbour, equal in altitude and otherwise similar, may have none. Those on a tract of high ground several miles in length may be confined to a space less than a quarter od a mile in diameter. There is still another factor. There is a consensus of opinion among those who have made burial mounds a special study, that the prehistoric people, especially of the Bronze period, had a preference for burying their dead on summits and brows, and there is no reason to doubt this.

That many of the tumuli occur in groups is well indicated on our map. In consequence, however, of its small scale, they are unavoidably exaggerated in size. For instance, there is a compact group of thirty-three on Cefn Merthyr, 4 miles S.S.E of Merthyr, which according to the scale would be three miles long, whereas it actually covers about one-fifth by one-tenth of a mile—a plot that would scarcely be noticed on the map. This group is about 1,212ft. above the Ordnance datum, overlooking the valley of the Taff. A mile to the S.E., near the opposite brow of the hill, is another, but rather scattered, group. On Gellygaer Common, to the N.E. of this, is a group of thirteen ascending to 1,570 ft. and covering about three-quarters by one-half of a mile. Several of these cairns have been rifled, and their cists left exposed. On Carn y Wiwer, near Pontypridd, is a compact group of thirteen at a height of 1,142ft., and on Hirwaun Common is another of twelve about 1,600ft.; while on Mynydd Caerau are nine, reaching a height of 1,823 ft. In the "Vale" between Cowbridge and Nash Point, are several small groups (all on relatively high ground), the most notable being eleven at Llanfrynach; and on Barry Island are, or were, six. On Grey Hill, Llanfair Discoed, are many tumuli, some kerbed with stone slabs on end. ("Arch." Camb.," 1909, p. 277.)

As already intimated, there is reason to think that the majority, at least, of these "other" tumuli belong to the Bronze period. An early mound of the period in Merthyr Mawr Warren was excavated by Mr. W. Riley in 1904. It consisted of a core of sand, covered with a thick crust of stones. On the old natural surface were two cists containing human skeletons unaccompanied by any grave-goods. At higher levels, and simply buried in the sandy core, were three more skeletons, each with a beaker or drinking-cup. All the skeletons were brachycephalic or broad-headed, and were laid in the contracted or flexed posture almost invariable in prehistoric unburnt burials. ("Arch. Camb." 1919, p. 336.) To the same period must be assigned a neighbouring tumulus of precisely similar construction, which yielded upon excavation in 1901, four cists containing as many broad-headed skeletons, but without any grave-goods ("Arch. Camb." 1919, p. 330); also two interments in cists accidentally discovered in 1902, the one near New-house Farm, St. Fagans, and the other at Cwm Cae, near Dolygaer. The former contained two broad-headed skeletons and a beaker, and the latter some fragments of decayed bones, a beaker, and a small flint arrow-head. ("Arch. Camb." 1902, pp. 25 and 28.) Another moundless cist at Llancaiach, near Gellygaer, containing a skeleton and a beaker, may be of somewhat later date, to judge from the shape and decoration of the beaker. During some excavations made by the Rhondda Naturalists' Society, on the site of a prehistoric settlement on Mynydd-y-Gelli, near Ystrad, in 1906, the fragments of a beaker were found among what appeared to be the remains of a destroyed cist and cairn. ("Arch. Camb." 1906, pp. 299 and 302.)

As the centuries of the Bronze period passed, cremation became more and more the vogue, and eventually seems to have entirely superseded simple inhumation in many parts of Great Britain. The earlier burnt remains are found associated with unburnt skeletons, but later, and during the food-vase period, they occur also as separate burials,

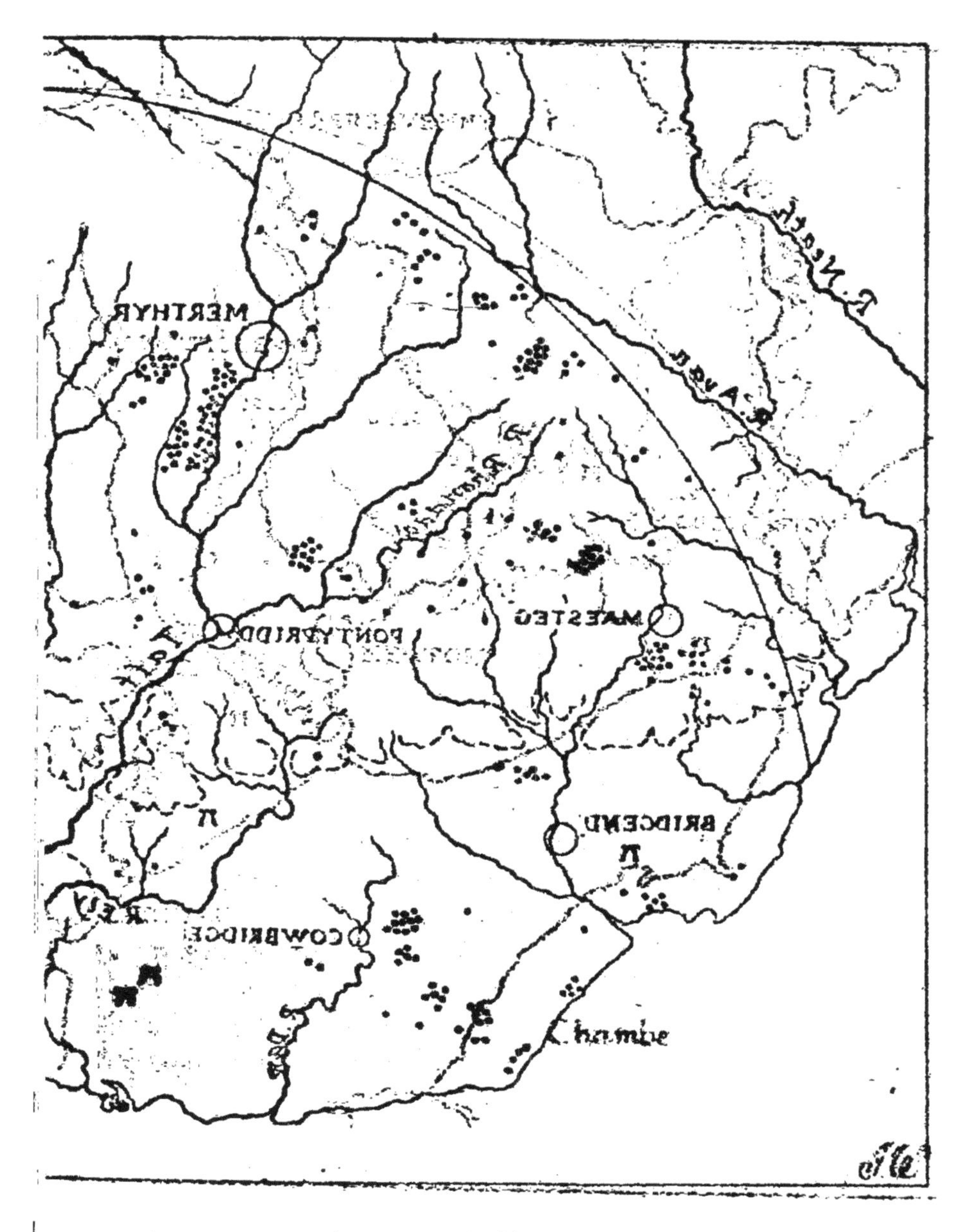

Map of the Cardiff D[illegible]

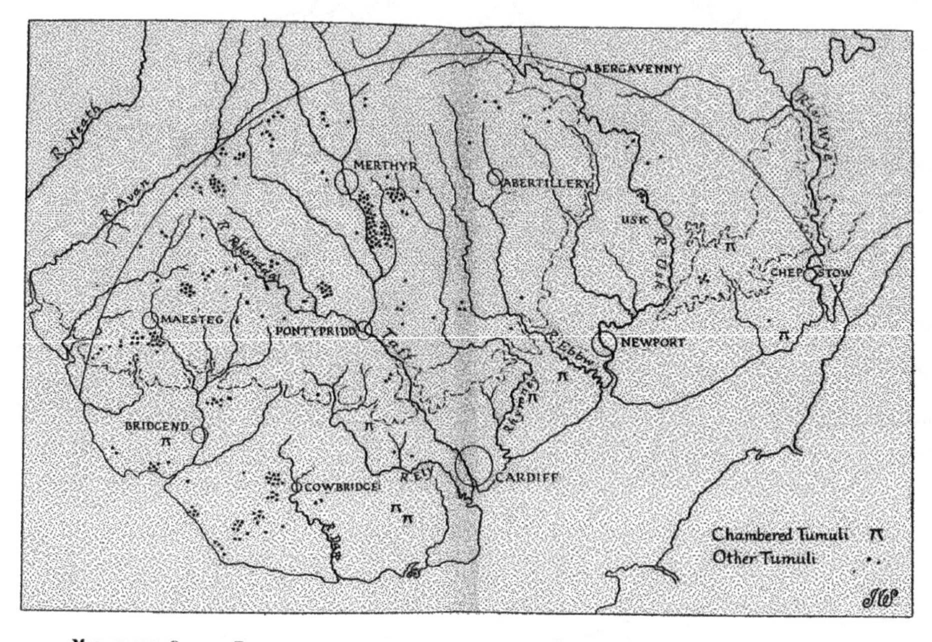

Map of the Cardiff District showing the distribution of the Prehistoric Tumuli. 8 miles to 1 inch.

sometimes in cists or in clay-lined holes, but more often apparently quite unprotected, although occasionally there are indications that the ashes had been deposited in a basket or a box, or wrapped in canvas or a skin. Occasionally a food-vase, *inter alia*, accompanies the remains. Later came the familiar cinerary urn which, unlike the latter vessel, is confined to burnt interments. These vessels are found throughout Great Britain, and in such numbers as to indicate that the custon of burial in them was both widespread and lasted for a considerable time.

The only discovery known to the writer of a food-vase interment in our district was in a ruined cairn near Candleston Castle, Merthyr Mawr, in 1901. It consisted of burnt bones accompanied with a rude food-vase, bronze dagger-knife, and small heap of snail-shells (*Helix memoralis*), all in a paved cist. ("Arch. Camb." 1919, p. 328.)

Of the six tumuli on Barry Island, one yielded in 1873, a burnt interment covered with an inverted cinerary urn, and three others without urns, but surrounded with sea shells and pebbles. ("Arch. Camb." 1873, p. 188; 1875, p. 185; and 1893, p. 71.) In 1894-5, Mr. J. Storrie cut wide trenches through all the mounds and found that all had been opened, but in one there remained part of a skeleton with three flint implements and a deposit of burnt bones with a chert implement, in a cleft of the rocky floor covered with a thin slab. Mr. Storrie learned that some time before, the skeleton of a young female had been discovered in a cist in another mound. ("Notes on Excavations made during 1894-5 at Barry Island," 1896.)

While removing stone for road mending from a small tumulus near Llysworney in 1865, a cinerary urn containing burnt bones was found. ("Arch. Camb." 1865, p. 207.) A remarkable tumulus known as the Twmpath, near Colwinston, was thoroughly explored under the direction of Mr. F. G. Hilton Price, F.S.A., in 1887. The central portion had been excavated several years previously by the then owner of the property, when nine cinerary urns and their contents were brought to light; but as these were not the sort of treasure he looked for, the work was abandoned. Mr. Price found that the tumulus contained a central space about 25 by 18ft., enclosed with a rude wall. He discovered nine more interments, all burnt. Of these, five were in cinerary urns (one inverted over the remains); one was in a small cist; and three were in holes lined with clay which had been hardened by fire. With these various interments were part of a bone pin, some fragments of bronze, and three bone beads. Of the eighteen interments found in this mound, fifteen were in the central enclosed space, and these Mr. Price regarded as contemporary and primary; he also mentions the tradition of a local battle, the slain of which were said to be buried here. His careful description, however, leaves little room for doubt that the burials took place at different times. ("Arch. Camb." 1888, p. 83.)

A large tumulus at Penhow, Monmouthshire, was opened in 1860, but the record of it is vague. An interment, whether unburnt or burnt is not stated, was found enclosed in a small (inner ?) mound and with it, or elsewhere in the mound, were two bronze dagger-knives, two flint flakes, and a "foreign stone." ("Arch. Camb." 1909, p. 278.) It may come as a surprise to many that there is some reason to think that Llandaff Cathedral is on the site of a pre-Christian cemetery. During some alterations in the nave in 1889 were found some deposits of charcoal and burnt bones, which had no apparent relation to the existing fabric. ("Arch. Camb." 1889, p. 266.) Mr. William Clarke, who was an eye-witness, confirms the discovery, and states that part of a bronze spear-head was found, but was mislaid or lost. Whether these deposits were as old as the Bronze period is uncertain, but it adds interest to the venerable fane to know that it may occupy a site accounted sacred long before the Christian missionaries appeared on the scene.

Circles (*Cromlechs*).—There is no certain example of a so-called druid's circle in our district. Supposed remains of them on the moors have been reported from time to time, but have not been accepted by those competent to judge. On Grey Hill, there are several small rings of stones, but there is little doubt that they are the kerbs of cairns which have been destroyed for the sake of their materials. There are in our district, as elsewhere in Wales, modern imitations of the prehistoric circles, some of which are liable to deceive the stranger. One is erected wherever the annual gorsedd is held, under the delusion that the ancient ones were of druidical origin and were raised for the same purpose.

Ancient Defensive Earthworks other than Roman.—Monmouthshire is remarkably rich in these remains, which are especially numerous in the vicinity of the Usk and of the Wye, and in less degree in the intervening portion of the county which comes within our circle. In Glamorgan they are mostly in the Vale and the contiguous parts of the hill country, especially between Margam and Maesteg; whereas in the more hilly north, they are few. Although they are generally of small size and the largest are only of moderate dimensions, they are thoroughly representative of their class at large, every type being represented, and for this reason the local examples offer an excellent field for study.

Two good examples of the promontory type in our district are Castle Ditches near Llantwit Major, and Caerau near Cardiff. The former is on the wedge of land between the sea and the Colhugh River at their confluence. With one side an almost vertical sea-cliff, and the other, a precipitous drop into a valley, the enclosure, which is about 5 acres, is completed with a formidable transverse line of entrenchments. On the face of the cliff is seen a section of the ditches, which were originally very deep, but now are half-filled with debris from their sides and from the rampart. This barrier stops short of the valley side so as to leave space for an entrance. The camp was originally larger, as the cliff is being eaten back by marine action. The other camp is larger and more complex, and is the largest fortified area in our district. being about 12 acres. It is on the western spur of a tract of high land, the sides of which are naturally steep, but have probably been rendered more so by man. There are no traces of an earthwork rampart or parapet along the brows. At a distance of 1350 ft. from the tip, the spur is indented with the hollow made by the water of a spring, and advantage of this has been taken in the construction of the artificial barrier, which seems to have been the work of more than one period. The entrance is in the centre, and on the north side of this is a ring of earthwork, apparently an addition to the original work.

The Bulwarks on the side of theWye south of Chepstow, is a good example of the 'cliff' type. The river bank here is precipitous, and this, with the artificial work, encloses an area of about 1½ acres. There are several small camps of the type along the coast, one at Sudbrook, near Caldicot, and two west of Wick, but they look like remnants of larger camps which have been partly destroyed by marine erosion. At Dunraven Castle are a number of entrenchments which apparently formed part of a large and important coast camp.

There are between one and two dozen camps in the district which thoroughly comply with the definition of the 'contour' type. An excellent example crowns Wilcrick Hill, an isolated hill near Caerleon. It is of oval shape, with several lines of entrenchments, which enclose about 3½ acres. Caerau, near Llantrisant, is equally imposing and is more than double the size. According to tradition, it is the site of battle fought in A.D. 873. The camp at Lodge Wood, Caerleon, is remarkable for its long, narrow shape (1,450 ft. by 450ft.) and its intricate multiple entrenchments. There are small but good examples of the type near St. Nicholas, and at Coed-y-Defaid and The Mount, near Bassaleg, and Maindee, near Newport. They are of rounded shape and enclose from half to one acre.

Perhaps a larger number of the local camps should be regarded as transitional between these and the promontory camps. Piercefield Camp, north of Chepstow, closely approximates to the latter type. It is finely seated on a spur 200ft. above the Wye, and is long and narrow—a shape determined by the site—and encloses about 9½ acres. The

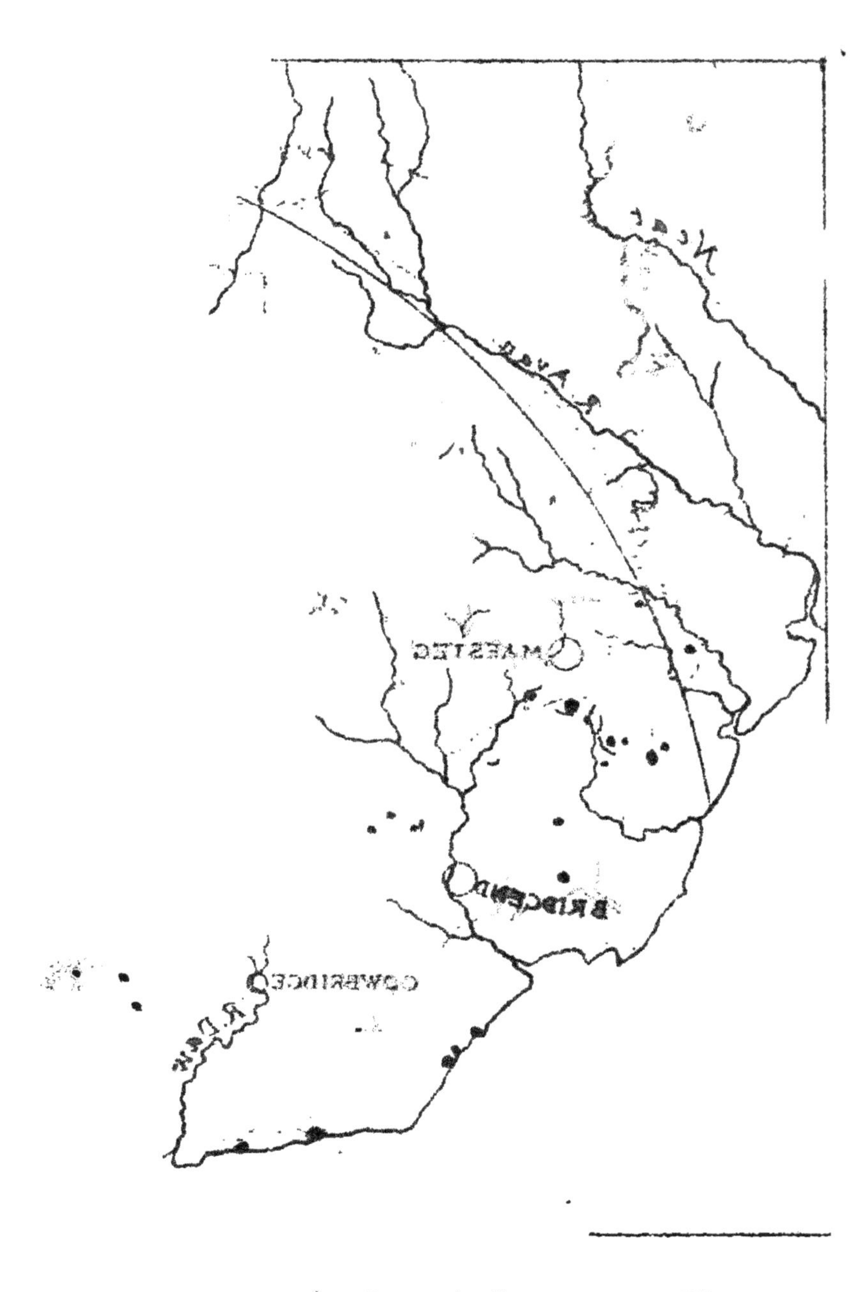

MAP OF THE CARDIFF DISTRICT SHOWING THE DISTRIB

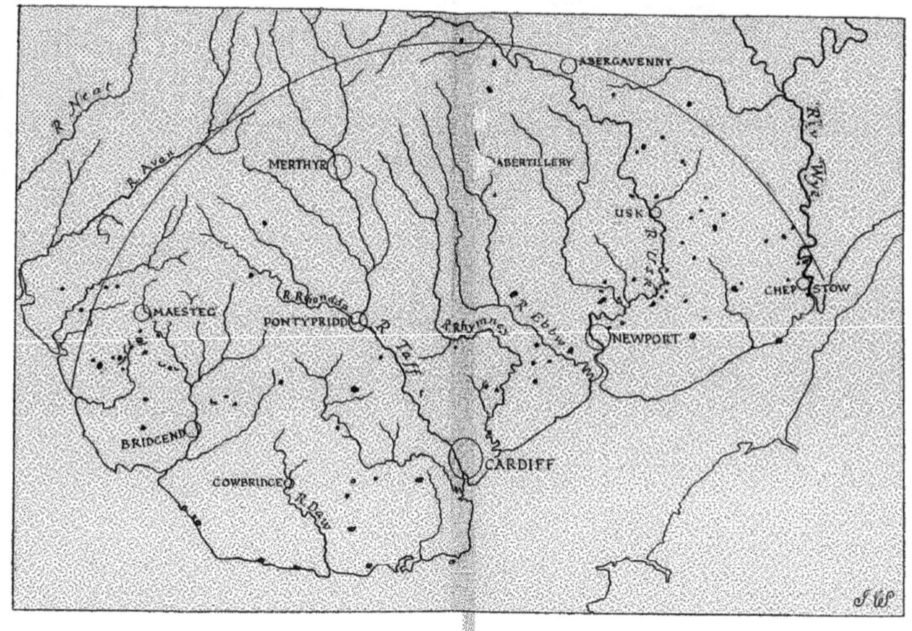

Map of the Cardiff District showing the distribution of the Prehistoric Camps. 8 miles to 1 inch.

west end, being the most assailable point, is strengthened by additional entrenchments. Pen-toppen-ash Camp, near Kemys Inferior, is a similar, but smaller example, irregularly oval and enclosing 4¾ acres. The camps which follow, lean more to the contour type. Castle Ditches, Llancarfan, is on a large spur of moderate elevation, and is one of the largest of the Glamorgan camps, being nearly nine acres in area. It is additionally intrenched where naturally most vulnerable. The camp known as the Fort, in Tredegar Park, near Newport, conspicuously crowns a spur-summit of considerable elevation. Its internal area is about 2½ acres, but its several lines of earthworks being widely spaced apart, cover a relatively large area of ground. At Coed-y-Bwnydd, near Bettws-Newydd, is a fine oval camp of 3½ acres, enclosed within several lines of entrenchment, with an entrance on the north-east, protected by a large mound. There is a similarly placed, but rather smaller, camp with several lines of intrenchment, in Llanmelin Wood, near Caerwent, which is peculiar in having an annexe containing three quadrangular divisions and a strongly fortified outpost a short distance away to the north-east.

A little camp cresting the lofty bank of the Rhymney at its highest point, near the village of Rumney, differs from a typical 'cliff' camp only in being entrenched all round. On Stormy Down, a stretch of high ground near Pyle, there is a camp of simple construction which certainly belongs to the 'plateau' type. It is almost circular and encloses about 2¼ acres. To the same type should be assigned a little triangular camp of two-thirds acre, on Mynydd Bychan, Llysworney; and probably another of similar size at Talgarth, Llanvaches. Besides the camps, of which many have not been noticed in this sketch, there are here and there in the district vague lines of entrenchment which appear to be relics of camps, which have otherwise been erased by agriculture or the slower ravages of time. Several camps have escaped the notice of the Ordnance Surveyors, notably one in the vicinity of Michaelston-le-Pit. ("Arch. Camb." 1913, p. 350.)

Of the relative ages and the uses of the ancient defensive works very little is known. There is no doubt that some date from Neolithic times, and there is a consensus of opinion that many, if not most, are legacies of the Bronze period, while a few were probably raised during the Roman conquest. Some of the smaller with slight fortifications, may have been thrown up during stress of war or the alarms of threatened invasion, but it is difficult to think that the greater and more intricate could have been undertaken under such conditions. More likely they were raised during peace, as a permanent provision for refuge when needed. Some had a permanent population and were practically fortified villages. Tre'r Ceiri and the fortress on Penmaenmawr in North Wales, and Whorlbury, Weston-super-Mare, are notable examples to the point. Whether any of the camps in our district contain the remains of huts can only be discovered by excavation, as almost all have been at one time or another under the plough.

On the Ordnance Survey most of the camps are named in black letter, which is the type used to indicate prehistoric remains; but in more than a dozen cases the names are in "sans," the type used for Roman remains, and in several instances (Caerau, near Cardiff, for one) the words ROMAN CAMP are added. It is difficult to understand why these camps should be attributed to the Romans, as they are entirely lacking in Roman characteristics and remains so far as is known.

The camps along the shore are popularly supposed to be the work of invaders, the invaders being Danes, Norsemen, or other sea-rovers. The camp on Sully Island (a peninsula at low water) is specified as a Danish camp on the Ordnance Survey. It is difficult to understand why these camps should be attributed to *invaders*, since there is no reason why the natives in the vicinities should not have availed themselves of strong positions on the shore. Apart from this, South Wales is a sinking area, and there is reason for thinking that the subsidence has been considerable since prehistoric times. Hence it is by no means certain whether any of these camps were on the shore at the time of their construction.

With the exception of one group, there does not appear to be any ground for thinking that the camps of the district or any considerable number of them, were the outcome of some general strategic scheme, or of concerted action on any large scale. The exception is a number of camps and lines of entrenchment on the uplands between Margam, Maesteg, and Bridgend, and especially on Mynydd Margam, which attains an elevation of 1,130 ft. These earthworks are of slight relief and for this reason many are partially obliterated. They have some other features in common, and a characteristic is their multiple trenches. On the eastern brow of the mountain are several camps and small posts which have been linked together by a line of entrenchment, and on the slopes to the south-east, also nearer Bridgend, are several small oblong posts. These various works appeal not only as parts of a general scheme, but as thrown up to meet a special emergency. They look like the field-works of an army and their disposition has suggested to some writers that they were raised to check an army advance from the south-east.

Huts and Settlements.—Although these must have been numerous, especially where the conditions were favourable for human habitation, very few discoveries of their remains have been reported in our district.

At the head of the Rhondda Valley, two miles north of Blaen-Rhondda Station, are the remains of a village of walled huts. They are scattered in several groups on as many terraces, on the south slope of the watershed between the river and the Neath. The walls are for the most part reduced to trails of stones. Many of the huts were circular with internal diameters ranging from 15 to 25ft.; others were irregularly rounded, and much larger, and possibly were enclosures for cattle. The western group is the largest, but most of the huts are in an extremely ruined condition. In 1913, Mr. Lleufer Thomas, M.A., had the site surveyed with a view to a spade investigation the following year, which, however, was not carried out in consequence of the late war.

On a similar terrace projecting from the side of the same valley near Ystrad, the Rhondda Naturalists' Society cut a number of rather inconclusive trenches on the site of what seems to be a walled village. The walls were about 8ft. thick, and to judge from the incomplete plan, there were several lines of them. The trenches brought to light hard earthen floors, burnt patches and stones, rude potsherds, stone pounders, flint implements and flakes, and apparently also the remains of a tumulus containing fragments of bone and a decayed bronze dagger-blade. (" Arch. Camb." 1902, p. 252.) In a similar situation 2 miles lower down the river, some faint indications of ancient remains led this society to do some digging with more definite results. The remains of two circular huts with internal diameters of 30 and 34ft., walls of several feet in thickness, floors of hardened earth, hearths, cooking-pits and scattered " pot-boilers," were brought to light. One of them had a small annexe containing a cooking-pit, and near by was an open-air hearth. Another small circular structure had a slab-lined trough extending the full width of the interior, 7ft. 6in., which contained dark earth at one end and a hearth on one side of it. A considerable number of small objects were found, some of which seem to indicate that the occupation extended into the Iron Age. (" Arch. Camb." 1906, p. 281.).

By the side of a small stream at Radyr, is an ancient cooking-place, which was first noticed by Mr. T. C. Cantrill, B.Sc., in 1910 (" Arch. Camb." 1911, p. 267), and was subsequently partially excavated by the Cardiff Naturalists' Society.

Rude Pillar Stones (*Maenhirs*).—No lapidary objects are more puzzling than these. Some are ancient memorials of the dead, and others are boundary stones, rubbing-posts for cattle, or gate-posts—also ancient perhaps, but more likely to have been erected within the last two or three centuries. Again, a stone set up for one purpose may now be serving quite another: a rubbing-post in a field may mark the grave of a long forgotten hero. On the moors of our district are many parish boundary stones, but while it is certain that most of them were set up for that purpose in modern times, some few, like the occasional prehistoric tumulus which has been utilised for the same purpose, are undoubtedly older than the boundaries they mark. It is probable that many of the modern stones were quarried, but there is no doubt that the majority (including all the ancient ones) were loose surface blocks when they were requisitioned by man; hence the weathered appearance of these is not necessarily due to their exposure since they were erected, and so is no safe indication of their antiquity as pillar-stones. But a peculiarity of the weathering may provide an indication in this direction. Rain-water falling on a stone collects in little streams, which take the lines of easiest fall; and if the stone remains long enough in the same position, the streams usually eat out little channels or gutters. If these channels are observed to follow the lines of easiest fall, the stone must have stood as it now stands for a very long time, many centuries at least. The size of a standing stone has sometimes also an inferential value. The well-known three monoliths of Trellech, Monmouthshire, are an instance to the point. Their heights above the ground vary from 8 to over 14ft. We have historical evidence that they are at least as old as the sixth century, for the village then bore its present name "Trellech" (three stones). ("Arch. Camb." 1861, p. 59.) There are standing stones near Llanvihangel Roggiett and on Grey Hill (several) in the same county, and near Margam (Gray, "Kenfig," p. 193), of heights varying from 6 to 8ft., which for the same reason may be regarded as memorial stones. Inference can be drawn from another source, not only as to the use, but also the age of the ancient stones of the class. There are about seventy inscribed stones in Wales, which, except for their inscriptions, would be indistinguishable from those we are describing. Their inscriptions are in debased Roman characters, with the occasional addition of oghams. The wording, which is in Latin and generally corrupt, indicates that they are grave-stones, and they are with general consent attributed to the 5th, 6th, and 7th centuries. Now it cannot be doubted that during those centuries the knowledge of letters was confined to very few, and it must have often happened that a grave-stone was set up in a district where there was no one capable of cutting or writing an inscription. It is reasonable therefore to think that many of our uninscribed examples belong to the Roman and post-Roman periods. Rude pillar-stones occur as "pointers" in connection with some prehistoric circles; but there does not appear to be satisfactory evidence of their use as grave-stones in pre-Roman times.

Cup-marked and other stones.—Near Gellygaer, there is a large weathered recumbent block of Pennant-grit, known as Maen Cattwg, the upper surface of which contains a number of cup-shaped depressions from 3 to 1½ inches in diameter. There is another recumbent block of the same rock on the summit of the west side of the valley at Bargoed, similarly pitted. In addition to the "cups" this has some faint grooves, but they may have been worn out by runnels of rain water. At Rhiwderin, near Rogerston, there is a third cup-marked stone which is figured in "Arch. Camb." 1895, p. 233. It is by no means certain whether the "cups" of these stones are artificial or natural.

Equally inconclusive as to whether it has any archaeological interest, is the Rocking Stone at Newbridge, near Pontypridd. According to Lewis's "Topographical Dictionary," 1843, it had been "much injured of late." As it is, with its additions, it is best described as a monument of Neo-Druidism. At Castellan, Llantrisant, is a natural ledge of rock, the sole claim of which to archaeological consideration lies in its popular designation—"The Altar."

Blown-sand Antiquities.—Along the littoral regions of South Wales, west of Cardiff, are stretches of land, mostly low-lying, which are covered with sand deposited by the prevailing westerly winds. Occasionally during storms, portions of the surface are left exposed through the shifting of the sand, and on these bared tracts may often be detected objects of various periods down to our own times, and the most frequent of these are flint implements, flakes and chippings. The most considerable stretch of blown sand in our district is the Merthyr Mawr Warren and its continuation, Newton Burrows. The first to call attention to the presence of worked flints here was the late Mr. R. H. Tiddeman, M.A., when engaged on the revision of the geological survey of the district in 1897. Since that time the Warren has been searched by others with considerable success, and many of the finds have been presented to the National Museum of Wales. ("Arch. Camb." 1919, pp. 324-5.) The flint implements are mostly small, and some well merit the designation of "pigmies." A notable feature is the stump of a polished stone axe from which flakes have been struck, and the small scatterings of chippings from other axes which have been found on several sites, but as yet no finished implement made therefrom has been reported. On the west side of Barry Island is a small tract of blown sand from which Mr. J. Storrie obtained a number of flint implements in 1894-5. (*loc. cit.*)

Cave Antiquities.—Two caves have been partially explored in the district—The one in Coed-y-Mwstwr, near Pencoed, by Mr. J. Storrie in 1884, and the other in the Little Garth, near Tongwynlais, by the late Mr. Trevor E. Lewis in 1912. The former yielded only a rough flint flake and a number of bones of animals of the historic period. The latter, although much blocked with fallen pieces of rock, yielded many Roman and post-Roman objects, and a few—as a nondescript piece of flint, portions of two vessels of coarse hand-made pottery, and a bone piercer and weaver's comb—which have some claim to be regarded as prehistoric. In 1908, five or six skeletons were found in a shallow recess in a steep slope of limestone in the Ifton Limestone Company's quarry near Severn Tunnel Junction. The bones had all the characteristics of early skeletons, and the skulls were markedly dolycephalic. The bodies had been laid in a contracted posture, and there were some indications that the recess had been walled in. ("Arch. Camb." 1909, p. 273, but a more detailed account was contributed by the writer to one of the Cardiff daily papers about May 13th, 1908.)

Hoards of Pre-Roman Bronzes.—Finds of these objects, singly or in twos, appear to have been rather numerous in the portion of Glamorgan that falls within our circle, but the recorded instances in the corresponding portion of Monmouthshire are few. There is a greater disparity in respect to *hoards*, for while six at least can be placed to the credit of the former, there does not appear to be one for the latter. When the railway was being cut through Coed Mawr at St. Fagans in 1849, a hoard was discovered, but there are no published particulars. In the British Museum there is a socketed axe from the hoard, and in the Welsh Museum four others and a fine socketed spear-head. (Evans, "Bronze Implements," p. 119.) In the Greenwell Collection are four flat axes which were found near Brithdir. (Evans, *op. cit* p. 43.) In 1887, a looped palstave, six socketed axes, axes, three socketed spear-heads, and a sickle were found in Colhugh Street, Llantwit Major. ("Arch. Camb." 1887, p. 151.)

More important was a hoard discovered in Llyn Fawr, near Hirwaun, in 1911. The tarn had been drained and was being prepared to serve as a reservoir for the water supply of the Rhondda Valley. In removing the thick deposit of mud and peaty matter, a number of bronzes were found, consisting of a fine large riveted cauldron with two handles, four axes, three gouges, and two sickles, all socketed; three convex discs, and four ornaments, apparently harness mounts. In addition, there were an iron socketed spear-head and sickle. The objects were scattered about a small space, as though they had been sunk from a boat. The iron spear-head was somewhat apart from the rest, but there is no reason to doubt its contemporaneity. No doubt whatever can be attached to the iron sickle, for it exactly reproduces in its form a socketted bronze one, and this must have involved much unnecessary labour in its production. It is, in fact, a clever example of the blacksmith's art, and we have in this hoard a most interesting legacy of the transition of the Bronze and Iron periods. The objects are in the Welsh Museum and an account of them will shortly appear in "Arch. Camb:" Another important hoard, also in the above Museum, consists of about thirty-six objects, many of which are of elegant forms and contain enamels. They were unearthed at Seven Sisters in the Dulais Valley, about forty-five years ago. They are typical examples of the art of the Late-Celtic stage of the Iron Period, and probably formed part of the hoard of a bronze-smith. Many of the ornamented pieces are harness mounts, and specially notable is a bronze weight, evidently the British pound. Altogether it is not only the most important find of the period in Wales, but some of the objects are unique in Great Britain. ("Arch. Camb." 1905, p. 127.) Probably belonging to a later stage of the Iron Period, were some bronze helmets and other objects found in a rock-cavity on Ogmore Down in 1818. They were all lost soon afterwards; but a short account of the discovery and a plate engraved from drawings made of them, were published in "Archaeologia,' XLIII., p. 553.

THE LORDSHIP OF GLAMORGAN.

By J. S. CORBETT.

THE following notes are intended to deal mainly with certain matters relating to that part of the old Welsh Kingdom of Morganwg which after the conquest, became the Lordship of Glamorgan, of which Cardiff was the head, and to give some information, necessarily of a general character, as to the government of that district up to the date of the Statute 27 Hen. VIII, cap. 26, which abolished the peculiar privileges of the marcher lords (or most of them) and introduced the English law and system of government. I do not deal with Wenllwch or Wentloog, now part of Monmouthshire, which (or a large part of it) formed part of the marcher lordship won by Fitzhamon, and remained so until the failure of the main line of the de Clares in 1314. To do so would unduly extend this article, and besides, it was treated in the de Clare s time as a separate "county."

PRE-NORMAN PERIOD.

As to the region to which this paper relates, the period from the end of the Roman occupation to the conquest by Fitzhamon is a very dark one, there being hardly any trustworthy authorities for its history. The "Brut y Tywysogion" and "Annales Cambriae" contain but very few and short references to this part of Wales. The so-called "Gwentian Brut" or "Book of Aberpergwm," claiming to be, as to its earlier portion, the work of Caradoc of Llancarvan (who is believed to have died about 1147), and the "Historia Cambriae" of Dr. David Powell (1584), purporting to give Caradoc's history, contain much relating to Glamorgan, but both must unfortunately be regarded as of very doubtful authority. The original or any MS. of Caradoc does not exist, and it seems only to be known from Dr. Powell's work. The "Gwentian Brut," as it exists, is probably not older than the

16th century, and though no doubt, to some extent founded upon older authorities, and perhaps in part upon Caradoc, it contains so much obvious fiction that, unless otherwise corroborated, its statements cannot be at all relied upon. The earlier part of the "Liber Landavensis" appears to the writer to deserve more attention than it has received, as it was actually written in the 12th or very early in the 13th century, much of it probably in the first half of the 12th. [In these notes references are to the edition by Mr. Gwenogvryn Evans, published in 1893.] Its authority has been depreciated (see Haddan and Stubbs "Councils, &c.") on the ground that the copies or records of charters and other documents of which it consists were put together, and perhaps to some extent altered or even fabricated, for the purpose of supporting the claim of Bishop Urban (1107-1133) to parts of the dioceses of St. David's and Hereford. That this objection has some force can hardly be doubted, but though it may be granted that the earlier records cannot safely be accepted as actual copies of documents of the dates to which they purport to relate, they show at least what was thought by a writer of the 12th century as to the succession of rulers of the district, and the statements as to facts of the 11th century, modern history at the time it was written, would probably be fairly accurate. It will be convenient, before going further, to refer to what the "Liber Landavensis" says as to the cantrevs contained, as it states, in the Kingdom of Morgannuc (Morganwg) and Diocese of Llandaff. The document occurs at p. 247, and is said to be inserted because the original was almost decayed from its great age. The first cantrev was Bican = Bachan (Carmarthenshire). The second cantrev Gwyr (Gower, now the part of Glamorganshire as constituted under Henry VIII. west of the River Neath). Cædweli (Kidwelly) and Carnwaliaun, both in Carmarthenshire.

These last three, in the "Red Book of Hergest," are treated as names of commotes making up a cantrev called Eginoc. The third cantrev was Wurhinit (Gorwenydd, now Groneath, Glamorganshire). The fourth, Penychen

(Glamorganshire). The fifth, Gunlyuch and Edelyvon (in Monmouthshire, the first name = Wenllwch or Wentloog, and the second Hedelegan or Edlogan, now a manor in Wenllwch). This last cantrev is between the Rhymney and Usk rivers. The sixth cantrev was Wenyscoyt (Gwent Iscoed or Underwood, the southern part of the land between the Usk and the Wye). The seventh was Wenthuccoyt (Gwent Uwchcoed or Overwood) and Ystradw (Brecknockshire) and Ewyas, always, it states, called the two sleeves of Wenthuccoyt, and Ergyn and Anergyn (Archenfeld, Herefordshire). This represents the Llandaff claim, not successful, as the diocese was ultimately settled so as to include only the Glamorgan territory east of Neath, what is now Monmouthshire, and one or two parishes in Herefordshire.

Whether or not the Kingdom of Morganwg at any time in fact included all these lands, the list is probably the oldest authority for the names of the cantrevs, and as such of value. It may be mentioned here that Giraldus Cambrensis, taking the St. David's view, describes the diocese of Llandaff as containing five cantrevs "and the fourth part of a cantrev, that is, Senghenydd." No doubt he meant to exclude the Carmarthenshire and Gower territory. Some authorities, the most important being the old list of cantrevs and commotes contained in the "Red Book of Hergest" (14th century) have mentioned another cantrev in Glamorgan called Brenhinol (Royal) or Breiniol (privileged); in the "Red Book" the form is Breinyawl. This, however, may be an error, possibly through attributing to a cantrev the epithet Breiniol, which, according to Rice Merrick (J. A. Corbett's edition, p. 119) applied in fact, not to a cantrev, but to the commote of Kibbor, formerly considered part of Senghenydd. This Senghenydd district, called by Giraldus the fourth part of a cantrev, would, according to the "Liber Landavensis" list, be included either in Penychen or Wenllwch. It lies in the eastern part of Glamorganshire, mainly between the Rhymney and Taff Rivers, but may perhaps at some remote period have been considered part of Wenllwch. From "Liber Landavensis" and the other authorities mentioned, supported in some points by certain statements of Asser, and the Anglo Saxon Chronicle, it appears clear that the territory which included the later Lordship of Glamorgan, was for some five centuries before the conquest, ruled by a line of princes descended from one Tewdric, slain in battle with the Saxons at a date perhaps not far from the year 600, and a contemporary of St. Teilo, after whom Llandaff Cathedral was named. These rulers are described at different times as Kings of Morganwg, Glamorgan, Gleuissicg or Gwent. These differences of description are perhaps due in part to the practice which often appears to have been followed of dividing a kingdom between sons of a deceased king. What exactly Gleuissicg meant is difficult to determine, but no doubt it included part at least of Glamorgan. It is suggested that it may possibly be an old name for Morganwg. Among the princes who sought the protection and friendship of Alfred, Asser mentions "Houil filius Ris rex Gleuising" and also "Brochmail et Fernail filii Mourici reges Gwent." Now Howel son of Rhys was a King of Morganwg, and in that character, according to "Liber Landavensis," made various grants to Llandaff. Brochmail and Fernuail were members of the same family. The Anglo-Saxon Chronicle states that Alfred's grandson Athelstan ruled over "Uwen King of the Gwentian people." Howel, son of Rhys had a son Owen, who may the person referred to, though here called King of the Gwentian people. Gwent, as appears above, consisted substantially of that part of the present Monmouthshire east of Usk. As to the names Morganwg (Morcanhuc, Morcannuc, &c., in "Liber Landavensis") and Glamorgan (Gulatmorcant), there seems to be a consensus of opinion that the former had a wider signification than the latter, though the names appear to mean much the same, "wg" being a suffix indicating a territory or district and "gwlad," from which the first syllable of Glamorgan comes, practically the same. An opinion has been held in some quarters that Glamorgan comes from "Glanmorgan," meaning 'along the coast,' but the fact that "Gulatmorcant" appears in "Liber Landavensis" in a document of the time of Bishop Joseph (early 11th century) supports the view that it means "Land of Morgan" from an early King of that name.

Space does not admit of saying more here as to the Welsh rulers of the district which includes Glamorgan, but it seems that (probably with some variations from time to time as to boundaries) their line ruled in practical independence of the princes of the rest of Wales until the 11th century, when a period of great confusion arose. It is stated in the "Brut y Tywysogion" that in 1021 Rhydderch son of Jestyn (not of the old Glamorgan line, but a descendant of Rhodri Mawr) assumed the government of the south, *i.e.*, Deheubarth, or South Wales. The "Liber Landavensis" (no doubt with exaggeration) describes him as reigning over nearly all Wales, and Howel of Glamorgan as "sub-regulus." Howel was of the old line of Glamorgan or Morganwg kings. Rhydderch was slain in battle in or about 1031. Much fighting took place between rival chieftains, and in the "Brut y Tywysogion" under 1047, it is stated that "all South Wales lay waste," though this may not have included Glamorgan, as the word is "Deheubarth" in the original. During his wars with the Welsh, Harold occupied Gwent, and built a hunting lodge in Gwent Iscoed, which was destroyed by King Caradoc, who seems to have been a grandson of Rhydderch. Then followed the Norman conquest of England, the appointment of William Fitz Osbern as Earl of Hereford, the building by him of Chepstow or Striguil Castle, and gradual encroachments on Welsh territory. It is clear that at the time of Domesday Book the Norman rule or overlordship extended as far as the River Usk and to some extent west of it, for a certain Turstin Fitz Rolf held some lands west of the Usk. The Welsh, however, were not completely expelled, and it appears that for a time at least Welsh kings, so called in "Liber Landavensis," ruled over at least parts of Gwent, no doubt in a tributary or vassal condition, and probably at times co-operated with the Normans against other Welshmen. For instance, the "Brut y Tywysogion," under 1070, says that Maredudd son of Owain was killed by Caradoc son of Gruffudd and the French (Normans) on the banks of the Rhymney, the river forming the eastern boundary of Glamorgan. In the "Liber Landavensis" (pp. 278-279) there is a statement that in the time of King William, Catgucaun (Cadwgan) son of King Mouric (who was son of the Howel before referred to) reigned in Glamorgan, and as far as the Towey (in Carmarthenshire), Caradoc in Ystratyu (Brecknockshire), Gwent Uwchcoed and Wenllwch, Riderch (Rhydderch) in Ewyas and Gwent Iscoed. This was no doubt the chronicler's statement as to what happened at some time during which things were in confusion and frequently changing. Most, or all of the Kings named were no doubt more or less under the domination of the Normans. In fact, it is said that all these Kings were under ("servierunt") King William, and died in his time. Caradoc was probably the same that was with the Normans in the fight on the banks of the Rhymney, but whether he was the same as the King Caradoc who, many years before, had destroyed Harold's hunting lodge, is a matter on which different opinions are held Rhydderch was a grandson o the Rhydderch slain in 1031. Cadwgan, as we have seen, was of the old line of Kings of Morganwg, but whether his rule in fact extended to the banks of the Towy is doubtful and when he died is not known.

Jestyn ap Gwrgan.

Later Jestyn ap Gwrgan appears as ruler (who is nowhere called King) of Glamorgan. Over what exact territory his authority extended cannot now be defined. He and his father Gwrgan are both mentioned in "Liber Landavensis" the latter (p. 263) as a witness to a document of the time of King Mouric son of Howel. Gwrgan and Jestyn are both mentioned in the Gwentian Brut as princes of Glamorgan, but the account given of them there is full of absurd and obvious errors and cannot be regarded in any way trustworthy. Still, there is no reason to doubt, though it cannot be said to be actually proved by any pedigree than can be fully trusted, that they were descended from the old line of Kings, and it seems probable that Jestyn assumed power on, or soon after, the death of Cadwgan son of Mouric.

FITZ HAMON'S CONQUEST.

In Jestyn's time, as is well known, the conquest by Robert Fitz Hamon took place. The legend so often repeated of his having called in Fitz Hamon to aid him against Rhys ap Tewdwr is more doubtful. In fact, it seems quite possible that Fitz Hamon's invasion does not represent the first Norman attempt upon Glamorgan. What the state of things was there at the time when Klng William made his pilgrimage to St. Davids (in 1079 according to the "Brut y Tywysogion," but more probably in 1080) is not known. The Brut, under date 1080, states that the building of Cardiff began. The Annals of Margam also (Rolls edition, p. 4) has, under 1081, "et edificata est villa Cardiviæ sub Willelmo primo rege." These statements no doubt point to the restoration of a fortress which had been desolated after Roman times, though as to when or by whom nothing is known. What this building or rebuilding may have led to, or what Fitz Hamon found at Cardiff on his arrival is at present uncertain. It was not until some 12 years after the dates mentioned as those of the alleged building that he defeated and expelled Jestyn. He had shortly before received from William Rufus the Honour of Gloucester, of which Bristol was the head, probably as a reward for his support against Bishop Odo and his confederates in 1088, and very likely with the intention that he should endeavour to conquer Glamorgan. As to the legend of his having fought with Rhys ap Tewdwr on the borders of Brecknockshire, it may be mentioned that the Brut and Annales Cambriæ say that Rhys was slain in 1091 by the French of Brecheiniauc, which seems much more probable than that Fitz Hamon should have penetrated so far into the hill country.

As to Cardiff, Fitz Hamon no doubt established himself there from the first, and made it the head of his Glamorgan lordship, but this does not prove that he found any town there. It may have commended itself to him on account of the remains of the Roman fortifications and also as a favourable place for communication by sea with Bristol. The place is not even named in any "Liber Landavensis" document prior to the conquest. As to the state of things in the district generally, there are no towns, castles, churches, or works of the kind of which remains exist which can be pointed to with any certainty as Welsh, and dating from the interval between the departure of the Romans and Fitz Hamon's conquest. This is no doubt due to the fact that the Welsh buildings were mostly of wood, for they certainly had many churches, besides their three important monasteries (possibly much decayed at the date of the conquest) at Llantwit, Llancarvan, and Llandough. It is true no doubt that some of their churches were of stone, but these, like Llandaff Cathedral, were rebuilt later, so that nothing seems to remain of early Welsh work, except indeed certain crosses or inscribed stones at Llantwit, Llandough, &c. It is to be supposed that the people lived mainly under the tribal system, described by Mr. Seebohm in his "Tribal System in Wales," and the late Sir John Rhys and Sir D. Brynmor Jones in their valuable work "The Welsh People." It is, however, likely that in the Vale district the tribal rules of descent may have given way to some extent even before Fitz Hamon's time, and the people have become more settled and more dependent upon agriculture than in the hills and wilder parts.

Another matter which has not been fully investigated is the question whether along the coast there may not have been settlements of Scandinavian, or at any rate other than Welsh people. All along the coast there are placenames of Scandinavian origin, and many English names, though, as to the latter, it is difficult to be sure whether they are earlier than the conquest, as no doubt many English came in afterwards. It seems certain that the Vale was subdued in a comparatively short time, and that the population settled down into the condition of manorial tenants under their Norman lords, for such disturbances as took place usually arose from incursions of Welshmen from the hills of Glamorgan or other parts of Wales rather than from local risings in the Vale. Though no doubt a good many Welsh chieftains were displaced, it may well be that to the actual tillers of the soil it made little difference whether they cultivated land for or paid their dues to a Welsh or a Norman chief or lord.

The Chief Lords.

Having reached the period of Fitz Hamon's conquest, it will be convenient to insert short particulars as to his successors down to the abolition of the marcher privileges and the application of English law, by the statute of Henry VIII.

The facts are briefly as follows :—

1107. Robert Fitz Hamon, djed.

Mabel, his daughter, married, at a date not exactly ascertained, Robert, a natural son of Henry I., created Earl of Gloucester and commonly known as Robert Consul (= Earl). 1147. Robert Consul died. The Countess Mabel died 1157. Their son William Earl of Gloucester succeeded his father. He married Hawise, daughter of Robert de Bellomont (called Bossu or Crouchback) Earl of Lincoln.

1183. (23rd November). Earl William died. His Countess survived and died 24th day of April, 1197.

Earl William had four children, Robert (who died young), Mabel, who married Almeric de Montfort Earl of Evecux, Amicia who married Richard de Clare, Earl of Hertford, and Isabel, who married John Earl of Mortain, afterwards King.

After the death of Earl William the Lordship remained in the hands of the Crown until, in 1189, King Richard I. made it over to his brother John, no interest in Glamorgan being allowed to Isabel's sisters.

John succeeded to the throne in 1199, and divorced Isabel in 1200, but retained the Lordship of Glamorgan until her remarriage in 1214 to Geoffrey de Mandeville, Earl of Exsex, who had it in her right.

1216. Geoffrey de Mandeville died.

1217 (before October). Isabel died childless. It is said by some writers that after the death of de Mandeville Isabel married Hubert de Burgh, but the better opinion appears to be that this was not so. Her lands were committed to him, but probably as custodian only.

1218. Gilbert de Clare, son of Isabel's sister Amica, succeeded. He was Earl of Gloucester and Hertford. He married Isabel, daughter of William Marshal, Earl of Pembroke, who, after de Clare's death, married Richard Earl of Cornwall, second son of King John.

1230. (25th October.) Earl Gilbert died in Brittany. His son, Richard de Clare, Earl of Gloucester and Hertford, succeeded. Born 4th August, 1222. Married Maud de Lacy, daughter of the Earl of Lincoln. On the failure of the male line of the Marshals, through his mother as one of the co-heirs, Richard obtained upon a partition of the Marshal estates considerable holdings in (amongst other places) what is now Monmouthshire. Through this and under certain subsequent arrangements, the de Clares had Usk, Caerleon, Trellech, and other lands.

1262. (July.) Richard de Clare died. The Countess survived him and died about 1289.

Gilbert, son of Richard, succeeded. Called "the Red Earl." He was born 2nd September, 1243. Builder of Caerphilly Castle. Married first, Alice de la Marche. He divorced her, and in 1290 married Joan, daughter of King Edward I.

1295. (7th December.) Gilbert de Clare died.

The Countess Joan succeeded for her life, in accordance with an arrangement made on their marriage, and in 1296, married Ralph de Monthermer, who during her lifetime, her son being a minor, sat in Parliament as Earl of Gloucester.

1306-7. (March.) The Countess Joan died.

Gilbert de Clare, the third of the name to hold the lordship of Glamorgan, son of Gilbert (2) and the Countess Joan, succeeded, Monthermer then ceasing to be styled Earl. Gilbert was born 1291, and married Maud, daughter of John, son of Richard de Burgh, Earl of Ulster.

1314. (June.) Gilbert de Clare (3) was killed at Bannockburn and the Countess Maud died 1315. They had one son, John, who died an infant in his parents' lifetime. With Gilbert (3) ended the male line of the de Clares, who had held the lordship from 1218 to 1314.

He left three sisters, daughters of Gilbert (2) and the Countess Joan. (1) Eleanor, of whom below. (2) Margaret, who married (1st) Piers Gaveston (2nd) Hugh d'Audley. (3) Elizabeth, amrried (1st) John son of John de Burgh, Earl of Ulster (2nd) Theobald Verdon, and (3rd) Roger d'Amory.

1317. A partition of the de Clare estates between the sisters was completed. Glamorgan fell to Eleanor, the eldest, who had married in 1306 Hugh le Despenser. son of Hugh, Earl of Winchester. Newport and Wenllwch, amongst other estates, were allotted to Margaret, and Usk, Caerleon, and other manors, &c., in what is now Monmouthshire, to Elizabeth.

Hugh le Despenser became Lord of Glamorgan in right of his wife Eleanor. He was the first lord who described himself in documents as "Lord of Glamorgan and Morgan," his predecessors having used the style "Earl of Gloucester and Hertford." He himself was not an Earl. He had two sons, Hugh and Edward.

1326. (18th November.) Hugh le Despenser was put to death at Hereford.

1328. William la Zouche married Eleanor, widow of Hugh, and after some difficulties on account of the marriage having taken place without the King's consent, became Lord of Glamorgan in her right.

1336-7. (1st March.) William la Zouche died.

1337. (30th June.) Eleanor died.

Hugh le Despenser (2) son of Hugh and Eleanor, succeeded. He was born probably in 1308, and married Elizabeth de Montacute, widow of Giles Lord Badlesmere.

1349. (8th February.) Hugh le Despenser died without issue. Edward le Despenser, nephew of Hugh, son of his brother Edward, who died before Hugh, succeeded.

He married Elizabeth, daughter of Lord Burghersh.

1375. (11th November.) Edward le Despenser died, his widow surviving. Thomas le Despenser, son of Edward, born 22nd September, 1373, succeeded. He married Constance, daughter of Edmund of Langley, Duke of York, who survived him, and died 1417. They had two children, Richard and Isabel. Thomas le Despenser was created Earl of Gloucester in 1397, but afterwards deprived of this title for conspiring against Henry IV.

1400. (January.) Thomas le Despenser, after taking part in an abortive rising, fled to Bristol, where he was beheaded by a mob, without trial.

Richard le Despenser, son of Thomas, died 13th October, 1414, while a ward of the King.

Isabel la Despenser, sister of Richard, succeeded. She was born after her father's death, 24th July, 1400, and married (1st) 27th July, 1411, Richard Beauchamp, 4th Earl of Worcester. The Earl of Worcester having died in France in March or April, 1422, Isabel married (2nd) 26th November, 1423, Richard Beauchamp, Earl of Warwick, a cousin of her first husband. They had two children, Henry and Anne.

1439. (30th April.) Richard, Earl of Warwick, died at Rouen, and Isabel died 26th December of the same year.

Henry Beauchamp, their son, succeeded. Born 22nd March, 1425. Married Cecilia, daughter of Richard, Earl of Salisbury. Her brother, Richard Neville, married Henry Beauchamp's sister Anne. Henry was created Duke of Warwick 5th April, 1444.

1446. Henry, Duke of Warwick, died, leaving an infant daughter, Anne Beauchamp.

1449. Anne, the infant daughter of the Duke of Warwick, died. Anne Beauchamp, sister of Duke Henry, succeeded her niece. She married as above mentioned, Richard Neville, who became Earl of Salisbury and Warwick (the "Kingmaker"), best known as Earl of Warwick, who held the lordship of Glamorgan in her right.

1471. (14th April.) The Earl of Warwick slain at the battle of Barnet.

The Earl of Warwick left two daughters, Isabel, who married George, Duke of Clarence, at Calais, 11th July, 1469, and Anne, who married (1st) Edward, Prince of Wales, son of Henry VI., and (2nd) in 1472, Richard, Duke of Gloucester, afterwards Richard III. On the death of the Earl of Warwick, the Duke of Clarence (ignoring the rights of his mother-in-law) entered upon the Lordship of Glamorgan. His brother Richard, on his marriage with Anne, demanded a share of the Warwick Estates, and an Act of Parliament was passed in 1474 to the effect that the Dukes of Clarence and Gloucester and their wives should have the estates as if the Countess of Warwick were dead, and might make partition of them.

1478. (11th March.) The Duke of Clarence was put to death. The Duchess Isabel had died 22nd December, 1476.

Either on the death of the Duke of Clarence or somewhat earlier, Richard, Duke of Gloucester, became lord, and later became King.

1485. (22nd August.) Richard was slain at Bosworth. His wife Anne had died previously.

Henry VII. succeeded to the Crown, and on 2nd March, 1486, granted the lordship of Glamorgan to his uncle Jasper Tudor, Duke of Bedford. However, Anne, Countess of Warwick, widow of "the Kingmaker" who was still living, had been *de jure* the person entitled to the lordship since 1471, but for the Act of Parliament before referred to, and it appears to have been desired to recognise her right while securing the Duke of Bedford's title. She accordingly petitioned Parliament in 1487 to repeal the Act which gave her estates to her daughters and their husbands, and this was done. She then, by deed of 13th December, 1487, in consideration of an annuity, granted the lordship of Glamorgan, &c., to the King. By letters patent of 21st March, 1488, he again granted it to the Duke of Bedford.

1495. (21st December.) Jasper, Duke of Bedford, died childless. King Henry VII, then held the lordship until his death. (There had been an Act of Parliament of 1496 that the possessions of the Duke of Bedford should go to Henry, second son of the King, but practically the King had control, his son being a minor.).

1509. (21st April.) Henry VII. died.

Henry VIII. succeeded. Up to the time of the Act about to be mentioned, he was styled in documents relating to Glamorgan "King of England and Lord of Glamorgan," and had a Chancery at Cardiff, &c., as had been the case under the earlier lords. The Statute of 27 Hen. VIII., cap. 26, for assimilating the laws and government of Wales and the Marches to those of England, which is further referred to later, appears practically to have put an end to the old Lordship of Glamorgan, abolishing the peculiar marcher jurisdictions and *jura regalia* while preserving the various lords' feudal rights with respect to lordships, boroughs and manors held under them.

The Knight's Fees.

To return to Fitz Hamon, there is no doubt that he divided the greater part of the Vale district amongst his followers, but we have but very little real evidence as to the arrangements made by him, for clearly the particulars given in the "Gwentian Brut" and Dr. Powell's history as to his twelve knights and their holdings are very far from accurate. Some of the names there given are doubtless those of followers of Fitz Hamon, but others are those of families who did not appear in the County for long afterwards, while others who probably were in fact among the earliest comers are omitted altogether. It is not until the "Liber Niger," 1166, that we have, in the carta or return of knights' fees held under William, Earl of Gloucester, grandson of Fitz Hamon, any authentic record as to the holders of manors in Glamorgan, and as they (or their predecessors) had for the most part held their fees from the time of Henry I., it is probable that they, or many of them, were descendants of the original conquerors. It is true that the late Mr. Clark, in his "Land of Morgan," p. 56, expresses the opinion that the return of the Earl of Gloucester did not include his Glamorgan fees, but his attention had doubtless not been called to the evidence afforded by a comparison of the "Liber Niger" with an Extent of Glamorgan (in the Public Record Office), which was certainly made in or about 1262, and in all probability on the occasion of the death of Richard de Clare, in that year. While the "Liber Niger" gives the names of holders and the number of fees held by each, it does not state where the fees were situate, but the "Extent of 1262" (as it is convenient to term it) does give this inforamtion, and it is the case that in that "Extent" many families of the same names as occur in the "Liber Niger" are recorded as holding in Glamorgan the same number of fees as are stated in "Liber Niger" to have been held by persons of those names. In 1262 the number of old fees, apart from some the holders of which are described as "noviter feoffati," is 36½ and ¼, paying wardsilver or castle guard silver to Cardiff Castle at the rate of 6s. 8d. per fee. This payment was not made by the "noviter feoffati."

The following is a comparison of the "Extent of 1262" with "Liber Niger":—

EXTENT, 1262.			LIBER NIGER.		
Name.	Place.	No. of Fees.	Name.	No. of Fees.	
Robertus de Someri	Dinas Powis	3½	Adam de Sumeri	7	Held 3½ fees elsewhere than in Glamorgan.
Johannes de Cogan	Cogan ..	2	Milo de Cogan ..	2	
Heres Gilberti de Costantin	Costantinestun (Cosmeston)	1	Robertus de Constantino	1	
Walterus de Sulye	Suyle }	2			
	Wenvoe }	2			

EXTENT, 1262.			LIBER NIGER.		
Name.	Place.	No. of Fees.	Name.	No. of Fees.	
Walterus de Gloucestria	Wrencheston (Wrinston)	½			
Willielmus le Soor	In Sancto Fagano	1	Jordanus Sorus	15	14 of these in Gloucestershire.
Willielmus Corbet	In Sancto Nicholao	3			
Gilbertus Umfravile	Penmarc ..	4	Gilbertus de Umfravill	9	5 fees in Devonshire
Willielmus de Kayrdif	Llanirid (Llantrithyd)	½	Willelmus de Cardi	½	
Phillippus de Nereberd	Aberthawe (St. Tathan)	4	Willelmus de Nerbertone	4	
Adam Walensis	Landochhe (Llandough, near Cowbridge)	1	Filius Ricardi Walensis	1	
Johannes le Norreis	Penthlin (Penllyn)	2	Defeodo quod fuit Roberti Norensis	2	
Willielmus de Wincestre	Landau (Llandow)	1	Roger de Wintonia	1	
Walterus de Sulye	Lanmais }	⅔			
Heres Gilberti de Costantin	.. }	⅓			
Thomas de Haweye	In Sancto Donato	1			
Quod heres Ricardi le Butiler tenere debet	Marcros ..	1	Lucas Pincerna Regis	1	
Hawisia de Londino	Uggemor (Ogmore)	4	Willelmus de Londonia	4	
Daniel Siward	Merthur Mawr	1			
Adam de Pireton	Nova Villa (Newton Nottage)	¼			
Abbas de Morgan (Margam)	Langewy ..	1			
		36½ & ¼		47½	
Deduct total in Glamorgan identified in Liber Niger ..		25	Deduct from these as not being in Glamorganshire— De Someri 3½, Le Soor 14, Umfravile 5	22½	
There remain in Extent of 1262 but not identified in Liber Niger		11½ & ¼	The total identified in Glamorgan is	25	

N.B.—The wardsilver at 6s. 8d. for each fee amounted to £12 5s. 0d. and this remained the same, and was paid in respect of the same fees throughout mediaeval times, with some small exceptions easily accounted for.

These $11\frac{1}{2}$ and $\frac{1}{4}$ fees in Extent of 1262, but not identified in Liber Niger, are made up thus (the names being those in the Extent of 1262).

Sully	4
Corbet, St. Nicholas	3
Llanmaes	1
De Haweye, St. Donats	1
De Gioucestria, Wrencheston	$\frac{1}{2}$
De Pireton, Newton	$\frac{1}{4}$
Siward, Merthyr Mawr	1
Abbot of Margam, Langewy	1

On this list of fees and holders appearing in the " Extent of 1262 " but not in the " Liber Niger, " the following observations may be made.

The de Sully holding probably is in fact in " Liber Niger," but in some other name. The family surname of de Sully may not have become fixed at that date. The statement that Sully was held as two fees seems to be an error in the original. Other inquisitions, &c., make the de Sullys hold one fee in Sully, one in Coychurch, and two in Wenvoe. The fee in Coychurch seems ultimately to have gone to the Turbervills, with whom there was controversy about it in the time of King John. Though this matter is one of some difficulty, it is clear that Sully was one fee, not two.

As to St. Nicholas, which William Corbet is said in the Extent to have held as three fees, it is there added " et tenentur in feodo de eo," so that he had evidently parted with the whole by way of sub-infeudation. It does not appear that a Corbet was at any time a resident landowner. Who, at the time of " Liber Niger," held the St. Nicholas fees, has not been ascertained.

The same is the case as to the one fee of Llanmaes held in 1262, $\frac{2}{3}$ by de Sully and $\frac{1}{3}$ by Costantin.

As to St. Donats, held by De Haweye in 1262, it may perhaps not have been granted at the date of " Liber Niger," or, if granted, was then in some other name.

The same remark applies to Merthyrmawr and Wrencheston. Newton Nottage was granted to a Sanford by William Earl of Gloucester, after " Liber Niger," it would seem.

Langewy or Llangewydd was in the late 12th century held by the Scurlags who made it over to Margam Abbey.

It affords the only instance in Glamorgan of a Knight's fee being held by an Abbey. Isabel, Countess of Gloucester (d. 1217) remitted all services except the payment of wardsilver. The name Scurlag does not appear in "Liber Niger," so it may be that the grant of Llangewydd to them was subsequent to it.

The Extent next states that Gilbert Turbervill held the Honour of Coytif (Coyty) by serjeanty of hunting. Also that Elias Basset held ½ fee in St. Hilary, and Philippus de Nereberd ¼ fee in Llancovian (Llanquian). As to these two last there is some difficulty, taking this Extent alone, as, while they did not pay wardsilver, they are not reckoned among the "Noviter feoffati." The explanation appears to be that they were originally held, not of the Castle of Cardiff, but of the Lordship of Llanbleiddian. This appears probable from later documents.

The Extent, then, under the head "Noviter feoffati," gives the following :—

Fulco de Santford, Lecwichehe (= Leckwith)	¼
Henricus de Sulye, Pentirech (= Pentyrch)	¼
Willelmus Scurlag, Llanharry	¼
Gilbertus Turbervill (Newcastle)	1/10
Rogerus de Clifford, Kenefeic (= Kenfig)	½
Willelmus Mayloc, In capella (meaning Llystalybont)	½

Then as to Welsh holders—

Morganus Vochan (Vachan)	Half a cummod in Bagelan "per Walescariam" and does no service except a heriot of a horse and arms at death.
Duo filii Morgani ab Cadewalthan (Cadwallon)	Half a cummod in Glinrotheni (Glynrhondda).
Griffid ab Rees	Two cummods in Seingeniht (Senghenydd).
Morediht (Meredith) ab Griffid	One cummod in Machhein (Machen).
	All as above, *i.e.*, no service except a heriot.

It is difficult to explain why Machen should be included, not being a member of Glamorgan, but of Wenllwch.

There are many sub-manors, held by sub-infeudation under certain of the Lords of the manors referred to, but it is impossible, in the space available, to enter into particulars of these.

Enough has been said to show that in Fitz Hamon's time, or very shortly after, nearly the whole vale country of Glamorgan was divided into Knight's fees, which were in fact manors of the same kind as those in England, as later inquisitions and ministers' accounts show. Fitz Hamon retained in his own possession the important Manor of Llantwit, usually called Boverton and Llantwit, which always remained in the hands of the Chief Lord, and also some lands near Cardiff, the full history of which would occupy too much space. He is also said to have founded Kenfig. It has been stated by some writers that he held Cowbridge and Llantrisant, but this seems very doubtful. It is more probable that the St. Quintins of Llanbleiddian held Cowbridge, and that Llantrisant was built and the borough founded by Richard de Clare.

Church Lands.

Fitz Hamon dealt with the estates of the Welsh monasteries by conferring upon English Abbeys the lands of Llantwit, Llancarfan, and Llandough, those of Llantwit and Llandough being given to Tewkesbury, and those of Llancarfan, or the greater part of them, to St. Peter's, Gloucester. These Welsh monasteries were no doubt very ancient foundations, the history of which cannot be discussed here, but, according to the chroniclers, they had been devastated on various occasions, particularly by "the pagans" in 987 ("Brut y Tywysogion") and it may well be that Fitz Hamon found them in a decayed state, and not, in his view, fulfilling any useful purpose. He also gave to Tewkesbury a church and lands at Cardiff. Probably he may have built the church or enabled the monks to do so, and he and his followers made them large grants of tithes at the expense of the Welsh clergy. With regard to Llantwit and Llancarfan, it seems possible that they had lost portions of their property before the conquest, for if the accounts of the colleges at those places given by early tradition are anything like accurate, those institutions must at one time have had much larger possessions than were conferred upon the English Abbeys. The Bishop of Llandaff continued to hold his Manor or Lordship of Llandaff with some special privileges and jurisdiction, and whether, or to what extent, Fitz Hamon despoiled the see is a doubtful matter, though other Normans in other parts than Glamorgan, are complained of as having done so, probably it seems by giving property claimed by the Bishop to monasteries founded or endowed by them.

The Comitatus.

The Vale of Glamorgan, after the conquest constituted the shire, or body of the County, administered by the sheriff, and the comitatus or County Court, regularly held monthly at Cardiff, but sometimes meeting at other times and places, and attended by the holders of fees held directly of the chief Lord and also by other leading men, holders of sub-manors, and also by some Welshmen, holding their lands in Welshery, as it was called, and not by feudal tenure.

The Member Lordships.

The shire proper, however, constituted only the smaller, though the more fertile and populous portion of Fitz Hamon's marcher lordship, for the greater area by far consisted of what were known as the "members" not considered part of the County until the Statute 27 Hen. VIII., cap. 26. These member lordships were Senghenydd, Miscin, Glynrhondda, Neath, Avan, Tir-yr-Iarll, Coyty, Ruthyn, Talyvan, and Llanbleiddian; Senghenydd being divided for some purposes into Senghenydd supra and Senghenydd subtus and Neath into Neath citra and Neath ultra. The last four, Coyty, Ruthyn, Talyvan, and Llanbleiddian, were under Norman lords, if not from the first, at any rate from very early times. These were held by Serjeanty and their lords had various independent privileges, holding perhaps under the chief lord a somewhat analogous position to that which a marcher Lord held under the King. It may be supposed that they held these special powers for the better defending their territories against the Welsh in the early days of the conquest. In fact, all the member lordships were often called marcher lordships (for instance by Rice Merrick) though that description, it is thought, could only properly apply to lordships held directly of the King. Tir-yr-Iarll appears to have always been considered as immediately under the chief lord, but in the earlier days his actual authority in the wilder portion of it must have been very small. The later lordship so called consisted and consists of the Parishes of Llangynwyd and Bettws, but prior to the gift of a large part to Margam Abbey, founded by Robert Consul in 1147, the name of Tir-y-Iarll is believed to have applied to a much larger district. In Llangynwyd at an early date, a castle was built, of which some remains exist, but as early as 1296, in the I.P.M. of Gilbert de Clare, it is described as burnt in war, and appears never to have been rebuilt. Tir-yr-Iarll was administered from Kenfig in the later times. Neath was granted at first to Richard de Granville but made over by him or his son to Neath Abbey, founded 1129, (or perhaps a little later), the chief lord, however, having a castle there, but neither de Granville nor the Abbot of Neath had any effective control over the Welsh of the hills which formed the northern part of the nominal lordship, and because he could not rule his Welshmen, the Abbot in 1289 exchanged it (excepting certain portions near the Abbey) with Gilbert de Clare, the second of the name who was Lord of Glamorgan, for an annuity or rent charge of £100 per annum, charged on the rents of certain boroughs and manors.

All the rest of the members, Senghenydd, Miscin, Glynrhondda, and Avan (sometimes referred to as Baglan) were in the hands of Welsh lords, who held under the chief lord, but probably really retained possession partly because of the barren nature of their country, and partly because they could not easily be expelled, and were no doubt practically independent. They were all (with perhaps some little doubt in the case of Senghenydd) descendants of Jestyn ap Gwrgan. There are, in the cases of all these Welsh lordships, indications of encroachments or attempts at such before their final annexation by the de Clares. Various abandoned mounds which probably have been "mottes" on which wooden castles existed (but which have not been fully explored) may represent either temporarily successful Norman attempts or erections of Welsh chiefs in imitation of the Norman practice. The status of such places as Whitchurch, Radyr, Pentyrch, and Clun, now parts of Senghenydd and Miscin respectively, but often referred to as Manors in mediaeval documents, seems to point in the same direction. Newland (the parish of Peterston super Montem) looks like an early encroachment by the lords of Coyty. It is on record that Ivor Bach, the Welsh lord of Senghenydd, surprised Cardiff Castle and took prisoner William, Earl of Gloucester, and compelled him to restore some lands which he had taken. (Giraldus Cambrensis and Annals of Margam, the latter giving 1158 as the date.) Where the seats of government, if the expression may be used, of the Welsh lords of Senghenydd, Miscin, and Glynrhondda were situate, is not ascertained, or how far they may have adopted Norman customs, but the Lords of Avan appear to have done so to a considerable extent, building a castle and founding the borough, now known as Aberavon, to which they granted charters, &c. As time went on it would seem that the position of these semi-independent lordships became intolerable to the chief lords. Their lords, whether Norman or Welsh, fought amongst themselves or rebelled against the authority of the chief lord, and in 1245-47, Richard de Clare expelled Richard Siward, then Lord of Llanbleiddian, Talyfan, and Ruthyn, and Howel ap Meredith, the Welsh lord of Miscin, and took those lordships into his own hands. At the time of the Extent of 1262 (as appears above) Senghenydd, Glynrhondda, and Avan (there called Baglan) remained Welsh, their lords, as is expressly recorded, owing no service except a horse and arms at death. However, Gilbert, son of Richard, followed up his father's policy, and in 1266 dispossessed and imprisoned Griffith ap Rhys, the last Welsh lord of Senghenydd, soon after building the great castle of Caerphilly, and probably also Castell Coch. He also, under some circumstances which do not appear to be recorded, obtained possession of Glynrhondda, for it is named as part of his possessions in the Inquisition of 1296 taken after his death in 1295. Neath, as we have seen, was acquired from the Abbey by this Gilbert in 1289. Avan remained in Welsh hands until the time of Edward le Despenser (d. 1375) who acquired it, it is stated, by exchange for other lands in England. Coyty alone never came into the hands of the Chief Lord, but its special privileges seem to have been curtailed. This, of course, was not a Welsh lordship, but was in the hands of the Turbervill family (probably followers of Fitz Hamon) and their successors.

Tenures.

As regards the Welsh lordships, we find in the Inquisitions hardly any such indications as are mentioned by Mr. Seebohm (in the "Tribal System in Wales") with respect to some parts of Wales, of a survival in full force of Welsh tribal customs. It rather seems as if the descent of land may have become simply gavelkind, though it is difficult to speak with confidence on the point. There are scarcely any traces of bond tenants, food rents, or servile works. All the tenures appear to have been free, or to have soon become so, except in a few cases in Whitchurch (a member of Senghenydd) and Radyr and Clun (members of Miscin), and it is highly probable that these had been annexed before the complete subjugation of those respective lordships. Otherwise there were practically no renders except small rents of assise, an aid called "comortha," and heriots at death. Very likely the rents of assise and comortha may have represented commutations already made in the time of the Welsh lords for former food rents. The comortha was a small payment by the owners of some (not all) tenements, made in most cases in every alternate year. It was peculiar to the hill lordships, and to Tal-y-van, in which latter place it seems to have been paid annually. There was also the avowry (advocaria), a payment of (ordinarily) 4d., by certain Welshmen described as holding no land, and also by sub-tenants, but this was not peculiar to the hills or to "member" lordships. It was received also by lords of manors in the Vale.

JURISDICTION.

The "member" lordships, whether originally held by Norman or Welsh lords, had, after the chief lord had taken possession, their own Courts with similar jurisdiction to the Comitatus or Shire Court at Cardiff. Nothing is known as to the manner of administering justice in earlier times, but the lordship courts under the chief lord were presided over by the Sheriff or a deputy, and in them fines of land were levied and all kinds of criminal cases tried. In cases of "false judgment," the suitors (who were the judges) might be, and sometimes were, fined in the comitatus, and so far their courts seem to have been in an inferior position, but we have no evidence that the fine went to the injured party. In the first instance, at least, it was accounted for to the lord. Rice Merrick, however, states that wrong judgments could be reversed by the Comitatus. The Inquisition on the death of Gilbert de Clare, slain at Bannockburn in 1314, states that each member lordship had "royal liberty" of itself, and also that Coyty and Avan were held with royal liberty, but that the chief lord had the rights of wardship and marriage, which is somewhat curious in the case of Avan, for lands held as it was called, "in Welshery," were not ordinarily subject to those incidents of tenure, and it seems as if some change had taken place there, if indeed the claim was admitted by the Lords of Avan.

THE WELSH 'PATRIA.'

Two other districts have to be mentioned, which were not portions of any manor, and though widely apart, were alike in one respect, viz., Kibbor and Glynogwr. In the case of Kibbor the Welsh portion is meant. Kibbor is, speaking generally, the district between the River Taff and Rhymney and the range of hills some five miles north of Cardiff on the north and the sea on the south. The north-eastern part was called the country (patria) of the Welshmen of Kibbor, who held their lands in Welshery, and were subject to no service except suit to the Comitatus at Cardiff. Glynogwr (Llandyfodwg) was in a similar position, except, perhaps, as to the suit to the shire court. It is now administered as part of Ogmore, belonging to the Duchy of Lancaster, but did not originally form part of that manor. It may be mentioned here that many manors contained small portions of land held in Welshery, owing no service except suit to the Court of the Manor. The farm known as Brynwell, in the Parish of Leckwith, is an instance, and there are several others in various places.

'Perquisites' of Court.

In those hill lordships which had long remained in Welsh hands, there was no great immigration of English, nor were the Welsh holders displaced. The chief instruments of oppression consisted of what were called the "perquisites" of court. These included heriots, and fines and forfeitures of every description. While these were and remained trifling in the boroughs and in the Vale, the inquisitions show that in the purely Welsh districts, they by degrees rose very greatly, and in Despenser times, must have been felt as a great hardship, very probably largely accounting for the support which Owen Glyndwr received there. The inquisitions show this to some extent, but they contain only estimates of the amounts. Court rolls or minister's accounts of the 14th century are almost entirely wanting, but there happens to be an account of the time of Edward le Despenser which shows that in 1373-4 no less than £144 7s. 4d. was levied under this head in Senghenydd subtus alone, the total of all receipts being £189 11s. 0d. In the I.P.M. of Edward le Despenser taken in the following year these pleas and perquisites are only estimated at £30 0s. 0d. Another custom which, in the later times at any rate, seems to have been almost, though not quite, peculiar to the hill lordships was the "mise," a payment made to each new lord on his succeeding to the lordship and collected in yearly instalments spread over five years. Its origin has not been satisfactorily explained. It has been said to have been made in consideration of the remission of fines, &c., due at the death of the preceding Lord, and it has also been thought to have been a kind of aid to the new lord towards paying his relief to the Crown. In the time of Elizabeth it was a fixed amount from each lordship liable to it. These were Senghenydd, Miscin, Glynrhondda, Tir-yr-Iarll, Ruthyn, Avan, Neath, and Llantwit Major. Cowbridge is also mentioned, which may mean Llanbleddian, not otherwise referred to. Rice Meyrick represents mises, in Hugh le Despenser's time, as having been payable from most of the Vale manors and gives a list of "ploughlands" subject thereto according to a survey of that period probably not now extant, which includes the greater part of the Vale manors, but if the mise was then in fact collected there, the practice seems to have ceased early. In the hill lordships, the mise was collected, at least in part, as late as 1758, but has now become obsolete. Another payment to the chief lord, in his own member lordships and boroughs, and also by inhabitants of manors held by others under him, was the chenee or cense, sometimes called "smoke silver," or the "toll of the pix," a tax upon inhabited houses or hearths. In member lordships, not in the chief lord's hands, the lords of those lordships appear to have received this. It was of very small amount, 2d. or 1d. (different lordships varying), and in later times often commuted for some fixed lump sum, for the whole lordship or borough.

In the vale manors, where there are copyholds and special customs as to descent of lands, in some cases the descent is borough English or to the youngest son (in Llantwit the youngest son by the first wife), and in other cases gavelkind, which appears to be Welsh, while borough

English may have been introduced from England after the Conquest. In the lordship of Coyty there are districts called Coyty Wallia with gavelkind and Coyty Anglia, with borough English. In the vale manors the customary tenants, and some of the freeholders, owed various works, such as ploughing, mowing, reaping, harvesting, fold making, weeding, &c., but by the early part of the 14th century it had become the practice for the most part to commute these services for money payments, and in some ministers' accounts of the 15th century, these payments are included with the chief rents or quit rents.

THE BOROUGHS.

The mediæval boroughs of Glamorgan were seven, Cardiff, Llantrisant, Cowbridge, Caerphilly, Kenfig, Aberavon (formerly Avan or Avene), and Neath. There can hardly be any doubt that all these were founded after the Conquest, and sprung up in connection with castles. In the case of all, except Caerphilly, some charters exist (not always in the possession of the borough authorities) but in no case does it appear that the first charter exists. The earliest known charter always treats a borough and burgesses as in existence, and some of them refer in terms to matters being done or carried on as of old. The mediaeval charters may be described as of two kinds, those granted by the Lords, comprising privileges throughout the lordship, and dealing with the government of the borough, its constitution, officers, courts, and their jurisdiction, &c., and others by various Kings of England, which in no way relate to internal affairs, but confer or confirm freedom from tolls, &c., throughout England. In the case of Aberavon the original earliest charter (not in existence) was no doubt granted by some Welsh lord. A charter of Leisan ap Morgan, living in the middle of the 13th century, grants to all the burgesses of his town of Avene all the liberties in that town which the burgesses of Kenfig had so far as he was able to grant them, and this was confirmed in 1350 by his descendant Thomas de Avene. These are the only existing charters granted by other than the chief lord, and illustrate the powers enjoyed by a lord of a member lordship. It may be added that Edward le Despenser who acquired Avan of Thomas de Avene, in the year 1373 granted a charter conferring on the burgesses of Avene further privileges, including the right to hold a fair. This document mentions the præpositus (portreeve), but contains no provisions as to his election, or anything as to the internal government of the town, from which it may be inferred that these matters had been already regulated by the Welsh lords, though the charters are not extant. It is impossible here to go into the details of the charters of the other boroughs. Some exist as to each, except Cowbridge (as to mediaeval times) and Caerphilly.

There is no doubt that Cardiff was the earliest in point of date, and the charters of other boroughs were more or less modelled upon those of Cardiff. The præpositus (portreeve) of Cardiff is mentioned in an Agreement of 1126 between Robert Consul and Urban, Bishop of Llandaff, and in a charter to Neath by Edward le Despenser there is mentioned as inspected a charter of William Earl of Gloucester (1147-1183), which granted to the burgesses certain privileges (as to freedom from toll) which the burgesses of Cardiff had. There is perhaps little doubt that some charter was given to Cardiff by Fitz Hamon. As to the boroughs generally, their chief municipal officers were portreeves, except (in the later mediaeval days) Cardiff and Cowbridge, in each of which two bailiffs were elected. The change from portreeves to bailiffs was made in each case in the 15th century, at Cardiff by a charter of Richard Beauchamp, Earl of Worcester, 1421, and at Cowbridge at some date between 1461 and 1487. The portreeves or bailiffs were elected by submitting certain names (three or four) to the constable of the castle of the town, out of whom he selected the portreeve or bailiffs. In each town there was held monthly or fortnightly what was termed the hundred court of the town, in which the constable, bailiffs (or one of them) or portreeve presided. The matters to be dealt with in this court occasion many provisions of the charters. The name "hundred" seems curious, for Glamorgan was not divided into hundreds until after, and in pursuance of, the Statute of 27 Hen. VIII., and the jurisdiction of the court did not extend to anything beyond the liberties of the borough. The charters prescribed the dates of fairs, contained provisions prohibiting trading by others than freemen, protecting the burgesses from being proceeded against (except in certain cases) elsewhere than in the town court, providing for the constable of the castle being *ex-officio* mayor, the making of bye-laws, and various other matters. The burgage rents varied in the different towns, being 1s. per burgage in some, as in Cardiff, for instance, and 6d. in others.

In the case of Cowbridge no charter is known to be extant (other than one of Charles II.), but a survey of the estates of the Earl of Pembroke made in 1570 contains a memorandum, not well composed or clearly expressed, as to certain charters said to have been granted by Richard de Clare, Hugh le Despenser, Edward le Despenser, Thomas le Despenser, Isabel Countess of Worcester, Richard Nevill, Earl of Warwick, and George, Duke of Clarence. It is not stated whether these charters then existed or where the particulars were taken from.

As to Caerphilly, nothing is known as to any charter, but presumably it had such, as it clearly was a borough, the latest, no doubt, to be founded, and the first to lose or disuse whatever privileges it may have had. It may be presumed to have arisen with the building of the castle (*c.* 1268). Each burgess paid 6d. for his burgage, and a like amount for an acre of land. It had a portreeve, hundred court, &c., as the other boroughs, but its privileges appear to have been disused during the 15th century. A minister's account for Caerphilly, 1428-9 shows that there was then a portreeve, serjeants, and borough court. In an account of arrears due to Richard, Earl of Warwick, in 1461, the portreeve of Caerphilly is mentioned as an accounting officer. But in a minister's account of the time of Jasper, Duke of Bedford, 1491-2, it is said that there was then no portreeve or serjeants and no court, the suitors doing their suit at the court of Senghenydd supra and subtus. It may be of interest to mention that in the case of the other boroughs the system of government by bailiffs or portreeves went on until the 19th century, in the case of Cardiff and Neath until the passing of the Municipal Corporations Act, at Aberavon until 1861, when a charter was granted by Queen Victoria constituting it a Municipal borough, and in the case of Kenfig, Llantrisant, and Cowbridge, until shortly after the passing of Sir Charles Dilke's Act in 1883, when Cowbridge reeived a charter as a Municipal borough, and Town Trusts were constituted for Llantrisant and Kenfig.

Officers of the Lordship.

The officers of the Lordship of Glamorgan were the Sheriff (sometimes termed the Sheriff of Cardiff, the name of the "caput" being then used for the lordship), who was appointed by the chief lord, not yearly, but apparently during pleasure. The sheriffs were sometimes selected from among the landowners of the County, but more frequently from outside. The sheriff presided in the comitatus or county court, and also in the chief courts of the member lordships when in the chief lord's hands, and appears to have represented the chief lord in his absence for practically all purposes. The Coroner was elected by the suitors of the county court or comitatus, by their submitting three names, out of whom the lord or the sheriff selected one. Rice Merrick states, as one of the privileges of Kibbor, that the person chosen as coroner must possess some land therein. He does not know the origin or reason of this. The coroner and his officers attended the comitatus for the purpose of enforcing its judgments. They also attended at the more important fairs to keep order, and for the protection of persons going to and from them, probably a very necessary thing in the then lawless state of the district. He collected the castleward payments or wardsilver due from the lords of the ancient manors, and certain miscellaneous rents, &c., due to the chief lord, but not arising from any manor in his hands so as to be collected by a manorial officer. He also had to do in some cases with the custody and conveyance of felons, and the realization of their forfeited goods. Under him were four bailiffs, those of East Thawe, West Thawe, Kibbor, and Glynogwr, and also sub-bailiffs.

The Bedells were the chief accounting officers in the member lordships of Senghenydd, Miscin, Glynrhondda, Talyvan, Ruthyn, Tir-yr-Iarll, Neath, and Avan.

In Senghenydd, Miscin, Talyvan, and Neath, there were also other officers called Receivers of the Forest, who are first mentioned in the I.P.M. of Edward le Despenser. Probably this may have been an office first introduced in his time. In the boroughs, the bailiffs or portreeve accounted for the lord's dues, and in ordinary manors the reeve. In each lordship or manor there was also a steward, who presided at the manorial courts. In the 15th century another official appears, called an "appruator," whose main business seems to have been to arrange or supervise letting of land, it having become a regular practice to let desmesnes, or other lands which had fallen into the lord's hand for lives, terms of years, or from year to year. There was only one appruator, who acted in various manors. The officers accounted for the moneys received by them to a receiver at Cardiff, and in the later times, at least, an auditor was employed to audit all the accounts.

Such is a short and necessarily imperfect account of the distribution of the lands and mode of government, if it can be called such, in Glamorgan prior to the Statute of 27 Hen. VIII., and from what has been said as to the composition and jurisdiction of the county court, and those of the member lordships and the boroughs, and when it is remembered that any criminal had only to make his way into Brecknockshire, Gower, or beyond the Usk in order to be safe, it is easy to understand the words used by Rice Merrick in speaking of the state of things prior to that Statute. He says, p. 88, "how unorderly they were then governed—Life and Death, Lands and goods, subject to the pleasure of peculiar Lords. And how uncertain lawes, customes and usages, whereof some rested in memory and not written, were ministered, a great number that live at this day can well remember and testify."

There can be no doubt it was high time that the *jura regalia* of the marcher lords should be abolished.

Perhaps a word or two should be added as to certain names of districts within the limits of Glamorgan and not ordinarily reckoned as "members" which are specially mentioned in the Statute of Henry VIII. as being in future to form part of the new shire.

Tallygarney (=Talygarn) in the parishes of Llantrisant and Pendoylan, had always been in Welsh hands, and its lords are said to have had powers of life and death, and some measure of independence, but it was not ordinarily counted among the member lordships.

Llandaff, the lordship of the Bishop.

Llantwit, always in the chief lord's hands, though not accounted part of the body of the County, and sometimes referred to as a member.

Ogmore, was originally an ordinary manor held as four Knights' fees, but had long since come to the King (Henry IV.) as Duke of Lancaster, and formed in the time of Henry VIII. part of the Duchy lands.

Doubtless it was considered best, in order to avoid doubts or questions, to mention these expressly.

LOCAL PLACE-NAMES.

By H. J. RANDALL, LL.B.

A student of the place-names of Wales is faced with the difficulty that there are few adequate authorities on the subject. The modern scientific study of place-names may be said to have begun with Skeat's *Place Names of Cambridgeshire* (published in 1901); but no competent scholar has yet produced a work upon the place-names of Wales as a whole, or those of a County, or one of the ancient territorial divisions. There is a certain amount of scattered work available, some of it of great excellence; but the most solid contribution to our knowledge is contained in the notes to the Cymmrodorion Society's edition of George Owen's *Description of Pembrokeshire*, especially those by Mr. Egerton Phillimore. Unfortunately the index volume has not yet been published, and that makes it very difficult to search out any particular reference in it. Under these circumstances, all that can be offered upon the present occasion are a few random notes upon the more obvious influences that have moulded the local nomenclature.

The central fact in the history of the south-eastern districts of Wales is that the whole territority was a borderland. To ignore that feature in the interests of "patriotism," or for any other reason, is to make misunderstandings inevitable. In the Middle Ages it was the frontier territory between opponents in arms, with a boundary line constantly shifting according to the fortune of war. The visible remains of that continuous and prolonged struggle have caused it to be named "the country of castles." When the active warfare gradually ceased, the struggle went on between rival cultures, rival languages, and rival creeds. The captains and the kings have departed, but they left South Wales, as they made it, essentially a borderland.

The record of its chequered history is written plainly

upon the place-names of the district. Beside the native Cymry, the Roman, the Dane, the Englishman, the Norman, and perhaps the Fleming, have left their imprints upon the topographical nomenclature. It may be interesting to offer a few observations upon the general evidence of their several influences; but first, in a class by themselves must be placed the river names.

Some of these may be Celtic, some may preserve traces of other tongues, but of the greater part one can only say that their origin is wrapped in the mists of an immemorial antiquity. There has been a natural tendency to frame Celtic explanations for almost all of them, but these attempts are little more than guesswork. For example, a name like the *Neath* has cognate forms in lands where no Celtic language was ever spoken, and for it, therefore, a Celtic origin cannot be predicated. It is at least a pleasant speculation that numbers of the British river-names may be the only surviving memorials of some prehistoric tongues the very memory of which has vanished from the face of the earth.

In this connection it may be well to mention a prevalent heresy that has ever affected the orthography of the Ordnance Survey maps, viz., that a multitude of river names ended in the suffix *-wy*, and that those that do not must be made to do so. Numerous instances could be given of the deliberate alteration of names to suit the purposes of this theory, in defiance alike of ancient forms and local pronunciation. Against this tendency it is only necessary to quote the emphatic words of the late Sir John Rhys. (*Celtic Folklore*, Vol. II., p. 516.) 'Those who have discovered an independent Welsh appellative '*wy*,' meaning water, are not to be reasoned with. The Welsh 'wy' only means an egg, while the meaning of 'gwy' as a name for the Wye has still to be discovered.'

Of direct Roman survivals there are few, as is the case in other parts of Britain. The actual Roman names were probably in most cases merely Latinised forms of the native words. The stations given in the Antonine Itinerary, Leucarum, Nidum, Bomium, and Isca are probably all to be attributed to this principle. The Welsh 'Sarn' (a causeway) is usually a clue to the existence of a Roman road. Sarn Helen, interpreted as Sarn-ar-lleyn, the causeway of the legions, is the name borne by many of the Roman roads in the country. "Street," an equally certain indication of the same thing in England, occurs in the names Blue Street and Water Street, near the ancient borough of Kenfig, clearly in the immediate neighbourhood of the principal Roman highway.

But most interesting of all these indications is the suggestion that has been made (Johnston, *Place Names*, p. 67) as to the origin of the name of the great seaport in which the British Association will assemble this year. Leland, an inveterate exponent of the art of guessing at the meaning of place-names, interprets it as "fort on the Taff" (*Caerdaff*). This is both historically and linguistically improbable. None of the early forms of the name suggests anything like it, while "Caerdaff" would never have become Cardiff. The name of the River Taff is preserved quite correctly in the adjoining city of Llandaff. The suggestion is that it may mean "fort of Didius." Aulus Didius was undoubtedly governor of Britain from A.D. 52 to 57, and carried on the war against the Silures. He is said by Tacitus to have pushed forward a few forts. Archaeological evidence has proved beyond a doubt the existence of a Roman station on the site of Cardiff Castle, though it was not mentioned in the Antonine Itinerary. The conjecture is at least highly interesting, and it has this at least in its favour that it does not offend against historical facts or philological theory.

The next invaders after the departure of the Romans were the Vikings; for the Anglo-Saxon settlements never touched South Wales at all in the Old English period. As early as the year 795 the coast of Glamorgan was ravaged by the Northmen and from that date onwards their visits must have recurred frequently. This is pre-eminently a subject in which the study of place-names can be used to reinforce our knowledge of a period for which direct historical sources are scanty in the extreme. Fortunately the work is being done in a thorough manner by Dr. D. R. Paterson, of Cardiff. His first paper on the subject (*Arch. Camb*. Jan.-Apr., 1920) can claim to be the only thoroughly scientific and orderly study of any section of the local place-names that has yet appeared in print. For the details the reader must be referred to the paper itself, but it will not be out of place to mention two or three general points.

In the first place, the names of the coast line and the few adjacent islands are predominantly Scandinavian. It appears to be a fair conclusion to draw from this that the invasions must have been frequent and regular, because it is, on the face of it, a somewhat astonishing fact that a foreign invader should have been able to impose his own names in his own language upon the whole coast-line, and that not for a brief time, but permanently.

In the second place, there is strong evidence of regular *settlements* of the Northmen in the Vale of Glamorgan generally. It is especially strong round Cardiff, Llantwit Major, and Kenfig, and along the line of the old Roman road, but it is by no means confined to these localities. Dr. Paterson's evidence is quite a revelation in the extent of the Scandinavian influence that it demonstrates. No such lasting results could have been produced by even a series of casual invasions ; the facts point to permanent settlements of an extensive character.

Thirdly, we may note that the Vikings, like all the other invaders, confined their attentions to the Vale country ; there is no record of their settlements among the hills.

Passing from the Northmen who came by way of the sea to the Normans who came by way of the land, we find that their influence upon the local nomenclature was as slight in Glamorgan as it was in England generally. The number of men of actual Norman blood who settled in the country was probably small, for the process of conquest in its earlier stages amounted to little more than a displacement of the chieftains by the incomers. 'Beaupré'—an ancient mansion near Cowbridge, is undoubtedly a Norman name, but

Beauville, near St. Andrew's Major, which seems to be the same, is really a corrupt spelling 'regardless of French grammar' of *Bovill* (Paterson, *op. cit.*, page 59). For the rest one can only say that Norman personal names may have entered into the composition of some place-names, such as Cornelly and Candleston, near Bridgend, but there are few even of these instances in which the possibility of the name being Scandinavian rather than Norman can be definitely excluded.

The English influence is another matter altogether. All along the Vale of Glamorgan the Englishman has planted his footsteps in no uncertain manner. Multitudes of names could be cited that are as purely English as any in Dorset or Sussex, but they have certain local characteristics, particularly in the endings. '—ton' is exceedingly common, and it is interesting to note that it is often applied to a single farm, as well as to a hamlet or village. Mr. Johnston (*Place Names*, p. 59) states that the same thing is characteristic of West Somerset and Cornwall, both districts in which English came into contact with a Celtic language. For the present we must be content to mention the facts without attempting to draw any inference from it. There are several examples of '-bridge' but hardly any of '-ford' in the more anglicised parts of the county. This is rather curious, as 'rhyd' the Welsh equivalent of 'ford' occurs fairly frequently. If -ton is common, some of the other usual English suffixes such as -ley, -ham, and -ing are very infrequent. A mansion house at Llantwit Major is known as the Ham; the Leys occurs near Gileston, and close to Nash there is a farm called Sheeplays, while Stalling Down, near Cowbridge, seems to be the sole example of -ing.

One of the fundamental differences between Welsh and English place-names is the predominance of personal names in the latter, and their great rarity in the former. The town and village names of England "tell us over and over again with aggravating monotony, how that an Englishman's house was and is his castle." Welsh names, on the contrary, are almost always descriptive, hardly ever personal. The numerous names of villages which bear the name in some form or other of the patron saint of the church are not really exceptions to this rule. The name of the patron saint was the distinguishing mark of the particular church and was therefore descriptive; a process quite different to that of naming a place as the -ham or -ton of a particular individual.

Possibly it may be a result of this Celtic influence, but it is certainly the fact that the English names in Glamorgan are more descriptive and less personal than is commonly the case. There are plenty of examples of personal names of the approved pattern like *Gileston* and *Walterston* and so forth; but side by side with these are an unusually large proportion of names in English of a descriptive character, such as Southerndown, Moorlands, Middlecross, Claypit, Broadland, etc.

The last of the invading influences was that of the Flemings. The alleged Flemish settlements in South Pembrokeshire and Western Gower are matters of historical controversy, and the whole problem deserves detailed examination by a competent scholar. All that can be said meanwhile, as far as South Glamorgan is concerned, is that there exists a group of names which merits proper investigation. They may turn out to be merely personal or family names, but on the other hand they may point to some degree of Flemish settlement. Examples are Flemingston, Flanders Farm, Flemings Down, and Pant Mary Flanders.

Finally we have to consider the predominant influence of the Celtic names. In order to do so a short excursion into local geography is necessary. The county of Glamorgan, like ancient Gaul, is naturally divided into three parts. The western peninsula of Gower is a separate region, outside the old Morganwg, and still for ecclesiastical purposes in the diocese of St. Davids. The rest of the County falls into two very well marked divisions, the boundary between which is roughly indicated by the Great Western main line from Cardiff to Port Talbot. The northern part, called in Welsh the Blaenau or hill country, is an old plateau at an elevation of about 1,500 feet above sea level, intersected by valleys that are always deep and usually narrow. Geologically it consists of the Pennant grit series of the coal measures. Needless to say it is the foundation of the modern industrial life of the county, and historically its importance rests upon the two features of its mineral wealth and its comparative agricultural infertility. It is a region of packed colliery villages and scattered hill farms with large sheep runs. In strong contrast to this is the Bro, or Vale of Glamorgan, claiming against some other competitors the title of the Garden of Wales. It is for the most part a limestone plateau ranging from 150 to 400 feet above sea level, and is a district of small farms, frequent villages, and high agricultural productivity.

The influence of these two regions is written in no uncertain manner upon the map of the place-names. In the Blaenau the native Cymru have all through the ages maintained an unquestioned predominance in language and culture. Whatever the political history of the district may have been, not one of the numerous invaders of Glamorgan ever seems to have succeeded in effecting a *settlement* in the hill country, or if such settlements were ever made, they have been so completely overwhelmed as to have left no trace of their existence in a local name. Village names, hill names, road names, field names, all are exclusively Celtic. One can say nothing more of this region than that it is one unbroken mass of Celtic influence.

Contrariwise, the Bro or vale country, the fertile and much to be desired prize of successful invasion, is a chequered mass of names of varied origin. It is this that makes the study of the region one of high historical interest. The bulk of the names even here are generally Celtic, but, so to speak, lying round this central core are the fossilised remains of the influence of the incomers. To examine the interesting problems that arise therefrom would be impossible in a limited space, but a few features may be briefly mentioned.

The strength of the Celtic core is demonstrated by the fact that, as far as the present writer has examined them, the *field* names are principally Welsh. Some purely English names occur, as well as some interesting examples of hybrid forms, but the foreigners are in a decided minority.

The character of a borderland is established by the occasional existence of two names for the same place. It is only possible to discover the fact by local inquiries, because the Ordnance Maps print the name that is in general use and give no indication of the alternative form. It is much to be hoped that a complete record of these substitutes could be compiled, as the remembrance of them is likely to disappear. Canon Bannister has shown in his *Place Names of Herefordshire* how between the date of the Domesday Survey and the end of the seventeenth century the Golden Valley was re-invaded by the Welsh, and the English and Norman names discarded in favour of Celtic ones, but in the Vale of Glamorgan the ultimate victory has more often rested with the English name.

Another borderland characteristic is the existence of hybrid names. Speaking generally, these are very rare. What usually occurs is that the *sound* of the original name is distorted into something that seems to resemble it in the other language, as in the well known instance of '*Yr afon glas*' becoming *Ravenglass*; or else the old name is entirely lost and a new one substituted. But in the Vale of Glamorgan, compound names, partly in one language and partly in another, sometimes occur. Tythegston, Coychurch, and perhaps Llanharry are examples, and the very curious farm name of Hamston Fawr.

Likewise, as Freeman pointed out long ago (*English Towns and Districts*, p. 7), names like Welsh St. Donats, or the two manors of Coity Anglia and Coity Wallia near Bridgend are specimens that one could only find in a frontier country. If it is necessary to distinguish one St. Donats from another by the fact that one is Welsh and the other is not, or to call one Coity Wallia and the other Anglia, it shows that we are upon the ground where peoples and languages meet. "If you are in Cardiganshire, you have no need to distinguish a place as Welsh Llanfihangel; if you are in Kent, you have no need to distinguish a place as English Dartford or English Sevenoaks." (Freeman, *op. cit.*, p. 8.)

Translations are another form of place-names found only in a borderland, and even there infrequently. Michaelston and Llanfihangel is a good example, and Bridgend has an exact Welsh equivalent in Penybont. In this last case the documentary evidence seems to point to the fact that the English form is the earlier, and the Welsh a translation of it. If so, it seems almost unique.

It is hoped that sufficient has been said to indicate the highly significant character of the place names of this ancient borderland, and to show that few districts would better repay systematic study by competent hands.

CARDIFF AND ITS MUNICIPAL GOVERNMENT.

By Mr. J. L. WHEATLEY,
Ex-Town Clerk.

The name and fame of Cardiff are perhaps more widely known than any other City in the United Kingdom, with the possible exception of London and Liverpool. The reason for this is found in that hidden treasure—*Coal.* South Wales Steam Coal is known throughout the wide world as "Cardiff Coal"—the principal Port from which this indispensable mineral is exported. This product proved a *sine qua non* in the great Armageddon which has now happily terminated.

As a Port, ancient history claims Cardiff as its offspring. The Charter bestowed in the reign of King Edward II. by Hugh le Despenser, Lord of Glamorgan (1359), granted to "our beloved burgesses of Cardiff" certain liberties in the nature of immunity from the payment of tolls, murrage, etc. From Lord Chief Justice Hale's tract entitled "De Portibus Maris" (written upwards of three centuries ago) it may be inferred that the Port of Cardiff was the most eminent on the north side of the Bristol Channel from Chepstow to Llanelly—the other Ports being "creeks" or "members" of the head port.

Cardiff is the only *civic* City in the Principality. Its municipal area is small, being 8,095 acres (including Flat Holm Island, 89 acres). In fact the area cannot accommodate its increasing population which is pouring over its civic limitations like the rush of water over a weir. Steps are now being taken to extend those civic limitations by the inclusion of a large additional area, a policy absolutely justified by its educational, industrial, commercial and political expansion. The present population is estimated to be 200,000. If the extension scheme is successful, the population will be considerably increased.

CHARTERS.

Cardiff has received some twenty charters in all, including a 12th century document. That of Incorporation was granted by James I., A.D. 1616. It comprised and extended those of Edward III. (1339) and (1359); Henry VI. (1455); and Elizabeth (1600). It became a County Borough under the Local Government Act, 1888. The Charter exalting Cardiff to the status of a City and its Mayor to that of Lord Mayor was granted by Edward VII. (1905). It is a Quarter Sessions Borough and an Assize Town. The rateable value for the year 1919 was £1,233,972.

GOVERNMENT.

The government of Cardiff is vested in a Council consisting of 40:—10 Aldermen and 30 Councillors. The City is divided into 10 Wards, three Councillors representing each Ward. It is proposed by the extension scheme that the City shall be represented by a larger number of Members on the City Council.

Up to the year 1918, it was represented in Parliament by one Member, and it was one of the largest single-member constituencies in the United Kingdom. This anomaly has been rectified by the Representation of the People Act, 1918, and the City, with the inclusion of the Urban District of Penarth (but excluding Cowbridge and Llantrisant) is now divided into three Parliamentary Divisions, viz., Central, East, and South, and represented by three Members of Parliament, a representation commensurate with the status and importance of the City and Port.

TRAMWAYS.

Cardiff affords signal illustration of the great development of municipal enterprise. Prior to the year 1898, it was served by horse-drawn trams, but under an Act of that year, the tramways undertaking was purchased from the Provincial Tramways Co., Ltd., at a total cost of £687,222. The tramways were immediately electrically equipped, and the system at present extends to all parts of the City.

Workmen's trams are run on all routes, and the Express Parcels Delivery system which has been established has proved a great convenience to traders, and is widely utilised. The Tramways Committee has never lost sight of the fact that to provide a cheap and rapid means of transit for the public is one of the most essential factors in the establishment of a Municipal Tramways Undertaking, and in this respect Cardiff is well served. It is proposed in the City Extension Scheme to extend and develop the tramways system and inaugurate motor omnibus service in the incorporated districts, a scheme which will prove of immense value to every section of the community. Several thousands of pounds have been contributed from the Tramways funds to the relief of the rates of the City, and for some years contributions out of the revenue of the Tramways have been granted in respect of the provision of music in the Parks.

The Power Station, main Car Sheds, etc., are situate at Roath, and cover an area of seven acres. The installation consists of four 300 K.W., two 900 K.W. Traction and Lighting sets, one 1,100 K.W. and three 2,000 K.W. extra high-tension three-phase sets for lighting purposes, one 100 K.W. steam exciter, and two 500 and two 300 K.W. motor generators. All the generators, with the exception of the 2,000 K.W. turbo sets are driven by slow speed vertical cross-compound engines. The boiler house contains 16 Lancashire boilers (each 30 feet long by 8 feet diameter) fitted with Vicar's mechanical stokers, and two water-tube boilers, also coal elevator and conveyor, and an ash elevator and conveyor for dealing mechanically with the fuel consumed.

The Corporation first embarked on the supply of electricity for lighting and power purposes in 1894, but since that date the plant installed at the generating station, situate at Eldon Road, has been added to considerably. This was necessary to meet the ever-increasing demand for current for power and lighting. Up-to-date machinery has been installed at this station and at the Central Sub-Station, The Hayes. A sub-station has also been erected in the Docks district.

WATERWORKS.

A detailed history of the Waterworks undertaking makes interesting reading. The main facts of it only can now be dealt with. In the year 1850 (when the population numbered about 18,000) an Inquiry under the Public Health Act of 1848 was held by Thomas Webster Rammell, Esq., into (amongst other matters) the water supply of the town. "The public supply of water for washing, cleansing and ordinary domestic purposes is obtained from the Canal, the River Taff, or from a few pumps in different parts of the town." Such was the condemning evidence of his report.

The chief supplies were derived from the public pumps in Crockherbtown (now Queen Street) and High Street, and were insufficient to meet moderate domestic requirements. Passing over intermediate stages in the history (during which a Company was promoted and reservoirs and other works constructed), we arrive at the period when the Corporation embarked upon one of the wisest projects to which they can lay claim in the annals of their civic history. In the year 1879 they acquired the Company's Undertaking, for which they paid £300,000. Then followed a prudent and far-seeing policy of expansion. In 1881 the Waterworks Engineer of the Corporation presented an exhaustive report to the Corporation and recommended what was known as the "Taff Fawr" Scheme, which included the construction of large storage reservoirs situate in the Brecon Beacons 30 miles distant from Cardiff, with trunk mains, etc., connected therewith. Concurrently with the obtaining of Parliamentary powers for this Scheme, the Corporation proceeded with the construction of a reservoir at Llanishen under powers granted to the quondam Waterworks Company. This Reservoir had a capacity of 317 million gallons, and was completed in 1886. While this work was being done, new and improved filter beds and service reservoirs were constructed at "The Heath." The Taff Fawr Scheme scheduled a watershed of over 10,400 acres, and the first reservoir (No. 2, or "Cantreff") was begun in 1886 and completed and opened in 1892—its capacity being 323 million gallons. No time was lost in proceeding with No. 1, or "Beacons" Reservoir, and by September, 1897, it was completed and opened. It has a capacity of 345 million gallons. Simultaneously with the construction of the "Cantreff" Reservoir was laid the "Taff Fawr Conduit" or pipe-line. Its terminals connect the Taff Vawr Reservoir with the Llanishen Reservoir. The conduit has a length of 32 miles, with subsidiary works *en route*. Some years elapsed between the completion of the "Beacons" Reservoir and the taking of definite steps to construct the third reservoir of the series. It was in the Parliamentary Session of 1909 that sanction was obtained for the construction of Reservoir No. 3 (or "Llwynon") and the work was commenced at once and proceeded with diligently when the spirit of military domination of a nation made it necessary to postpone its completion. This work and other additional and ancillary works, including the construction of a High Level Service Reservoir and a second pipe line from the Brecon Beacons to Cardiff are now being proceeded with, and when completed, Cardiff will have a total storage capacity of roughly 2,300 million gallons, sufficient for a population of over 300,000 persons.

The wisdom of embarking upon the Taff Fawr Scheme has been amply justified. The water from this source is very soft, containing from three to four degrees of hardness, and is of a high standard of bacterial purity. The Corporation supplies districts *outside* the City containing a population of approximately 44,000. There are a number of small Reservoirs, Filter Beds, and other works connected with the undertaking, which contribute to make the water supply one of the finest in the country. The Capital outlay up to the present time is approximately £1,674,133 6s. 10d.

Education.

In the cause of education (a subject dealt with in a separate article) the Corporation may claim to have shown a practical and liberal interest. Under the powers of the Cardiff Corporation Act, 1884, the Corporation subscribed £10,000 towards the funds of the University College of South Wales and Monmouthshire, and, pending the payment of this capital sum, paid to the College interest amounting to £1,623. In the year 1900, under powers conferred by the Cardiff Corporation Act, 1898, it made a free grant of a site of five acres, valued at £20,000, for the new College buildings in the Cathays Park, and in the year 1909 it extended the area by 1 rood 29 perches, at an estimated value of £1,725. The Corporation, as the successors of the Intermediate Education Governors, makes an annual payment of £800 on the condition of the provision of certain Scholarships. Up to the 31st March, 1916, the College, by agreement with the Corporation, provided accommodation for the Technical School, and appointed teachers in connection therewith, in consideration of payments amounting in 25 years to £106,606. These were of course payments for services rendered. The Corporation also makes to the College an annual grant of £1,050. A new Technical College has been erected in the Cathays Park, which, when fully equipped, will cost approximately £105,000. Other Educational Institutions, *e.g.*, Intermediate Schools, Pupil Teachers' School, Municipal Secondary Schools, and Council and Non-Provided Schools (numbering 37), bear testimony by their efficiency to the directive energies of the Corporation.

The Cardiff Corporation has not been unmindful of the ancient and illustrious past of the City. Evidence of their practical interest in this respect has been shown by the institution of a scholarly research into its history. In the year 1893 it established a Records Committee, and appointed an archivist of repute—the late Mr. John Hobson Matthews. After several years engaged in the examination and editing of documents, six handsome imperial octavo volumes, with illustrated and facsimile reproductions of Cardiff Charters were published.

Parks.

The Public Parks, Pleasure Grounds, and Open Spaces provided cover an area of over 300 acres, Roath Park alone accounting for 102 acres. This is a very popular park, which includes a lake of 32 acres, on which a motor launch plies and boating has its enthusiastic devotees. Fetes and galas and open-air concerts attract enormous crowds to this park, the land for which was presented to the Corporation by the late Lord Bute and other landowners upon certain conditions, which need not be specified here. The subjects of Parks must not be disposed of without a reference to Cathays Park of 59 acres, which —thanks to the wise and far-seeing policy of the Corporation—was purchased from Lord Bute in the year 1898 for the sum of £159,323. In this Park are located the City Hall and Law Courts, the University College of South Wales and Monmouthshire, the Glamorgan County Council Hall, the Technical College, the National Registry Offices of the University of Wales, and (now in course of erection) the National Museum of Wales, for the establishment of which last-mentioned Institution the Corporation fought with determination. The fight (which was long and strenuous and into the details and stages of which present space forbids entry was), unfortunately, only partially successful, as the sister institution (the National Library) was located at Aberystwyth. The Cathays Park is considered one of the finest sites for public buildings in Europe, and the series of national, civic, and educational buildings already erected and to be erected will constitute an achievement worthy of the most enlightened and progressive spirit of national and civic government.

Infectious Diseases Hospital.

The Corporation erected at a cost (including site) of £75,000 a Sanatorium or Hospital for Infectious Diseases. This Institution was opened in 1895; and its design was modelled upon the most up-to-date scientific principles and methods. The Corporation also possesses an Infectious Diseases Hospital on the Flat Holm Island, where cases of cholera and other allied diseases are treated. A crematorium has also been erected on the Island in connection with this Hospital.

Mental Hospital.

The City Mental Hospital which, during the period of the War, has been used as a War Hospital, was opened in April, 1908, and is built on a site of about 190 acres, situate some three miles from Cardiff, and provides accommodation at the present time for 750 patients, but with administrative departments adequate for 1,250. The total area of the building and airing courts is about 17 acres. The rest of the estate—deducting ornamental grounds—is used for farm and garden purposes. Space precludes any description of the institution. It is equipped much above the average in respect of scientific apparatus for clinical and pathological research. Gradually, as the result of research work into the causation and cure of insanity, a reputation is being (perhaps has been) established which will reflect constantly increasing credit upon the Visiting Committee who, under the presidency of Alderman Morgan Thomas, J.P., have exhibited an enlightened spirit and have been inspired by the ideal that these institutions should be regarded not as Homes of Rest or places of detention, but as institutions where scientific research is undertaken and fostered.

Baths.

The Corporation has provided swimming, Turkish, and ordinary baths and a gymnasium at Guildford Crescent. Both institutions are well patronised, and it is certain that the general health of the community is improved in consequence of their provision.

The markets provided by the Corporation are extensive, modern, and well-built, and are located in St. Mary Street, The Hayes, Roath, and Canton.

Streets and Bridges.

One of the essential contributory factors in the transformation of a town possessing the lineaments of antiquity to one of modern aspect is street improvements, and in this respect Cardiff has undergone a complete metamorphic process. As far back as 1877 the Corporation obtained power to construct new bridges over the River Taff and Glamorganshire Canal and new roads in connection therewith, by means of which better access was provided between the western and south western districts of Cardiff and the district of the Docks. Involved in this particular work was the construction of the "Clarence Bridge," which has a central swing span of 190 feet 8 inches.

Descending the chronological ladder we find the Corporation engaged on the laying out of the Cathays Park with its handsome King Edward VII.'s Avenue and other spacious thoroughfares. There remains one obvious public improvement still to be carried out, viz., Duke Street, which is the *bête noir* of every person with pretensions to an enlightened knowledge of the way modern towns should be planned. Had not the War intervened, the work would have been proceeded with and ere now well on its way to completion.

Closely allied in character to street improvements is the question of town planning and housing, in which the Corporation has taken and are taking a keen practical interest.

Sewerage.

The sewerage question in Cardiff, as in other Ports, has been a thorny and difficult one. The City's sewerage scheme is carried out on the "combined" system, and the sewage is discharged through three main outfalls in the Eastern, Central, and Western Districts. The main outfall sewer for the Western District discharges sewage into the open sea off Lavernock Point, about three miles distant from the nearest point of the City boundary. This particular outfall sewer cost £245,524 14s. 3d. to construct. Cardiff is low-lying, and the outfalls being necessarily below the level of high water, are consequently tide-locked for periods varying according to the height of the tide. The outfall sewers are therefore constructed to act as reservoirs for the storage of the impounded waters during the time that the outfall is closed.

The Corporation is the Burial Authority and owns a Cemetery of 95 acres.

Under the powers of the Small Holdings and Allotments Act, 1908, the Corporation purchased and leased land at Capel Llanillterne and at Llanedarne, and there established Small Holdings. A total acreage of 456 acres has been acquired for this purpose. In addition, some 432 acres have been obtained for allotments.

Police.

The City Police Force comprises a Chief Constable, a Deputy Chief Constable, 2 Superintendents, 2 Chief Inspectors and 12 other Inspectors, 38 Sergeants, and 236 Constables. The Headquarters of the Force are located in the Law Courts. There are in addition seven Police Sub-Stations in various parts of the City.

The Police Fire Brigade consists of a Chief Officer, who is the Chief Constable, 1 Superintendent, 1 Inspector, 3 Sergeants, 22 Police Firemen (full time), and 30 Police Firemen (auxiliaries). A new Fire Brigade Station has been erected in Westgate Street, and the Beasley-Gamewell fire-alarm system installed, as a cost of £27,293.

City Hall and Law Courts.

The City Hall and Law Courts (including the purchase of Cathays Park, etc.), cost £529,608 12s. 5d. The City Hall is a magnificent and imposing building. It is here that, through the wise generosity of the late The Right Hon. Lord Rhondda, M.A., the Welsh Historical statuary is located, perhaps the finest collection in the United Kingdom. It is here that most of the local government work is carried on, and where most of the great and epoch-making schemes of the future will be conceived, promoted and fostered.

THE PUBLIC BUILDINGS OF CARDIFF.

By W. S. PURCHON, M.A., A.R.I.B.A., Lecturer in Architecture and Civic Design in the Technical College, Cardiff, and HARRY FARR, F.L.A., City Librarian, Cardiff.

MODERN Cardiff is a city of late development, though it has a history which can be traced back to very remote times. Its rapid growth in recent times is due to the development of the economic resources of South Wales in the early nineteenth century. From a small provincial town of 1,870 inhabitants in 1801, it has become a metropolitan city with an estimated total population of 204,436.

As one would expect under such conditions of development, it has outgrown its public buildings more than once. Its Town Hall up to 1861 was in the middle of High Street. Afterwards a more ambitious building was erected in St. Mary Street, on the site of the new buildings of the Co-operative Wholesale Society. This, in its turn, has given place to the noble City Hall, now one of the Civic buildings in Cathays Park.

Before the City Council acquired Cathays Park, Cardiff could not lay claim to public buildings which would compare with those of other great towns in the United Kingdom, nor did it possess, apart from Cardiff Castle, mediæval buildings whose beauty and antiquity were likely to attract the visitor.

The purchase of Cathays Park from Lord Bute, however, gave the Corporation a site for public buildings right in the heart of the City, a site almost without a rival in Europe, and one which afforded a unique opportunity for grouping together a series of public buildings that might challenge comparison with those of other large towns. The visitor who, coming to Cardiff for the first time, makes his way to Cathays Park, cannot but admit that the opportunity has been well used.

Entering the Park from Park Place, the Law Courts on the left, the City Hall in the centre, and the uncompleted National Museum of Wales on the right at once come into view, and in their setting of trim lawns and green foliage, make a group of great beauty, an interesting touch being added by the old-world thatched building just beyond the Law Courts.

The City Hall and Law Courts.

Erected in 1904 from the designs of Messrs. Lanchester and Rickards, the City Hall and Law Courts form a dignified and graceful group which is generally admitted to show a striking advance on English Civic Architecture of the immediately preceding period. The design is based in the main on the best Renaissance work of France and our own country, but the architects have impressed their individuality on it in unmistakable fashion.

The two end pavilions on the main front of each of these buildings form settings for groups of symbolic sculpture. The group on the western pavilion is by Mr. D. McGill, and represents "Science and Education." The pavilions on each side of the avenue bear groups by Mr. Paul Montford, representing "Commerce and Industry," and "Music and Poetry," while on the eastern pavilion is a representation of "Welsh Unity and Patriotism," by Mr. Henry Poole.

The central portion of the City Hall is occupied by the portal, with its richly treated carriage porch, used on all ceremonial occasions. Above the porch is the central window of the Council Chamber, the domed roof of which is surmounted by a figure of the Welsh Dragon, modelled by Mr. H. C. Fehr, and cast in lead upon an unusually large scale. The two large groups of statuary flanking the centre window represent the sea receiving the three rivers of the city, the Taff, Rhymney, and Ely.

Perhaps the most striking feature of the City Hall is the Clock Tower which, rising high above the main structure, forms a magnificent landmark. The richly treated upper part of the tower, beautified by figures (modelled by Mr. H. C. Fehr) representing the four winds, gains much by its contrast with the severely plain sub-structure.

The other parts of the City Hall are designed in keeping with the main façade, but with greater simplicity. The breadth of treatment given to the sides by the plain masses of the end pavilions is particularly noteworthy.

The interior decoration is no less striking. The outstanding feature is the Marble Hall on the first floor, with its monolithic columns in Siena Marble, forming an approach to the Assembly Room of great dignity and beauty. The Assembly Room itself and the Council Chamber both contain much fine decorative work.

The City Hall has only been built for a comparatively short time, but already it has been enriched by many gifts of paintings and statuary. Two great gifts have been made. The first by Mrs. Andrew Fulton, widow of Alderman Andrew Fulton, J.P. (Mayor of Cardiff, 1884), who in 1907 bequeathed one-fourth of her residuary estate to be applied to the completion and decoration of the interior of the City Hall and the purchase of paintings, statuary, and the like. The other and more recent gift is that of the Welsh Historical Statuary by Lord Rhondda of Llanwern.

The paintings and statuary are nearly all on the first floor. On either side of the main entrance to the Assembly Room are two large pictures, one, "The Penitent's Return" by Sir Luke Fildes, R.A., purchased out of the funds of the Fulton Bequest, and the other a painting by Margaret Lindsay Williams of the National Ceremony of the Unveiling of the Welsh Historical Sculpture by the Right Hon. D. Lloyd George, presented by J. C. Meggitt, Esq., J.P., of Barry.

Near the entrance to the Council Chamber are two fine pictures purchased from the Fulton Fund, "The Shadow" by E. Blair Leighton, and "Winter" by Joseph Farquharson, A.R.A. Here also is the fine portrait by Herbert Draper of the late Lieut-Col. Lord Ninian Crichton-Stuart, M.P.

In the corridors leading off the Marble Hall are a number of paintings. In the corridor leading to the Luncheon Room are two, one presented by Mrs. Andrew Fulton, "The Holy Loch, Greenock," by James Greenless, and the other "The Bay of Naples," by John Glover, presented by John Griffith Jones, Esq. (Ap Caradog), of Pontypridd. In the corridor leading to the Lord Mayor's Room are portraits of Lt.-Col. Frank Hill Gaskell, of the 16th (Cardiff City) Battalion Welch Regiment, by L. De Berna, presented by his father, and of Sir W. J. Thomas, Kt., of Ynyshir, also painted by L. De Berna, presented by a few personal friends; two ceremonial paintings "The Ceremony of Investiture of the Lord Mayor of Cardiff with the Lord Mayoral Thumb Ring, November, 1911," by Frank Craig, presented by T. H. Riches, Esq., of Kitwells, Shenley, Herts., and "The Knighting of the Lord Mayor of Cardiff" (Alderman Sir W. S. Crossman, J.P.), by W. Hatherall; and an example of the work of a Welsh artist, Peury Williams, the "Procession returning from Festa of the Madonna del arco at Naples." The two latter paintings were acquired out of the Fulton Fund.

Farther along the corridor near the Town Clerk's office, is an example of the work of Mark Anthony (an artist who lived for a time at Cowbridge) "Through the Woods," and on the wall above the staircase leading to the Town Clerk's Office is a fine seascape by Norman Wilkinson, depicting "H.M.S. Cardiff leading in the surrendered German Fleet, November, 1918," presented by Lord Tredegar. On this landing is another example of the work of a local artist, "The King saluting the grave of an unknown Soldier," by G. F. Harris, presented by Sir James Herbert Cory, M.P.

In the Assembly Room are large portraits of General Botha, by Frank C. King, and of the Earl of Plymouth, while in the Council Chamber and Committee Rooms are many portraits of former Mayors and Lord Mayors in their robes of office.

In the Marble Hall itself are the unique series of statues of Welsh National Heroes, which the City owes to the munificence of the late Lord Rhondda of Llanwern. Each statue was entrusted to a gifted sculptor, with the result that the statues are not only representations of the most famous Welshmen, but are also fine examples of the work of the best contemporary British sculptors. The material used for the statues was the finest statuary marble, Serravezza, and the pedestals were made from Pentelicon marble with Siena marble panels.

The central figure is "Dewi Sant" (Saint David), the Welsh patron saint, by Sir W. Goscombe John, R.A. Flanking the staircase on the one side are pedestal statues of Llewelyn ein llyw olaf (the last prince), by Henry Pegram, A.R.A.; Harri Tewdwr (Henry VII.), by E. C. Gillick; Owain Glyn Dwr (Owen Glendower), by Alfred Turner, R.B.S.; Sir Thomas Picton, by T. Mewburn Crook, R.B.S.; and in the niche, Williams Pantycelyn, the great Welsh hymn writer, by L. S. Merrifield.

On the other side are Hywel Dda (Howell the Good), by F. W. Pomeroy, A.R.A.; Buddug (Boadicea), by J. Havard Thomas; Dafydd ap Gwilym, by W. Wheatley Wagstaff; Giraldus Cambrensis, by Henry Poole, R.B.S., and in the niche, Bishop Morgan (the first translator of the Bible into Welsh), by T. J. Clapperton.

Two other sculptural memorials will be found on the landings of the staircases leading to the Marble Hall. These are large bronze tablets with portraits in low relief, one commemorating Sir Edward J. Reed, for some years M.P. for Cardiff, by Sir George Frampton, R.A., and the other commemorating Capt. Scott, the Antarctic Explorer, who sailed in the *Terra Nova* from Cardiff on his last voyage, by W. Wheatley Wagstaff.

The Law Courts.

The architectural treatment of the Law Courts was skilfully devised in perfect harmony with the City Hall, but on somewhat severer lines. The main entrance front to the Avenue, with its fine double loggia, and the ably treated façade to the canal, both thoroughly merit the high praise they have received from many students of architecture.

The interior of the Law Courts is again fittingly treated more severely, but no less beautifully than that of the City Hall.

The National Museum of Wales.

To the east of the City Hall is the National Museum of Wales, now in course of construction. The portion already erected is only about a quarter of the complete scheme, as may be seen from the interesting model in the City Hall.

The Architects, Messrs. Smith & Brewer, have designed a building with an able plan and a noble architectural treatment worthy of its position as a national monument in a group mainly devoted to City and County purposes. The exterior, while refined to a marked degree, is by no means lacking in strength, and it seems clear that the structure, when completed, will possess that crowning quality of unity which is so rare in buildings of such dimensions.

Of the four groups of sculpture on the south front, the western pair, representing the Prehistoric and Classic Periods respectively, are by Gilbert Bayes, while the eastern groups, executed by Richard Garbe, represent the Mediæval and Modern Periods. The two groups on the western façade, by Thomas J. Clapperton, represent Mining and Shipping, are remarkable for their extraordinarily high architectural quality.

Of the interior, it is as yet too early to write, but those who have been fortunate enough to inspect it have been surprised and impressed by its magnitude and, even in its present incomplete condition, by its dignity.

The University Registry.

Adjoining the Law Courts on the west side of King Edward VII. Avenue is the University Registry, the first building to be erected on the Cathays Park site. The architects, Messrs. Wills and Anderson, designed this building in the maturer English Renaissance manner, and succeeded in investing it with a quiet academic dignity befitting its purpose.

The Glamorgan County Hall.

The next building in line with the Registry is the Glamorgan County Hall, a worthy companion to the Civic buildings of the Cardiff Corporation, though differing widely from them in conception. Its fine entrance hall leads to a noble council chamber, arranged to seat 88 members, beyond which is an interesting series of Committee rooms overlooking the North Road, while in other parts of the building may be found the offices of the various County administrative departments. The main façade, treated with a fine portico of coupled Corinthian columns, flanked by pavilions designed with an ability which shows most obviously in their lower portions, is worthy of comparison with any similarly classical composition; while the back elevation, with its balcony supported on well designed stone brackets, its curved wing walls, and delightful central feature, is remarkable for its grace and charm. This elevation alone would put its designers, Messrs. Harris and Moodie, in the front rank of English architects; they risked much in departing from the safe academic path, but in so doing they achieved sheer beauty.

It may be noted that all the furniture and fittings in this building and in the City Hall and Law Courts were designed by the Architects, a fact which largely contributes to the harmony of the interiors.

The large sculptured groups in front of the main façade of the County Hall were executed by Mr. Hodge.

The Technical College.

Beyond the County Hall comes the Technical College of the City, which was designed by Messrs. Ivor Jones and Percy Thomas, of Cardiff, and opened by the Lord Mayor in 1916. The architects relied for effect on beauty of architectural form and proportion rather than on ornament and sculpture, and its ably designed Neo-Grec façade is very dignified. The building, which is an excellent solution of a difficult practical problem, is in the form of a hollow rectangle, the entrance front and the two sides being completed, while the back block is in process of construction. Internally the main hall, which occupies a considerable portion of the interior of the rectangle, is a fine example of the excellent results which can be obtained inexpensively by a skilful designer.

The University College of South Wales and Monmouthshire.

Facing the centre of the Park on the east side are the buildings of the University College, which were opened by the President, The Right Hon. The Earl of Plymouth, on October 14th, 1909, while the Viriamu Jones Laboratory was formally opened on June 26th, 1912, by His Majesty King George V. The buildings were designed by Mr. W. D. Caröe in his well known manner—a version of the Earlier English Renaissance—and when the full scheme has been completed it will form a symmetrical group round a great central court. The Library, the gift of the Draper's Company, is in the form of a great galleried hall, and is one of the principal rooms in the portion of the building at present completed.

Public Statuary.

With the development of the Cathays Park has come the erection of some notable public memorials. On the green in front of the Law Courts is a statue of Judge Gwilym Williams, by Sir W. Goscombe John, R.A. In the centre, approaching King Edward VII. Avenue, is a memorial "To the memory of the Welshmen who fell in South Africa, 1899-1902," by Albert Toft. A list of nearly two hundred names follows. The panels bear bronzes with the names of the principal victories in which the Welsh Regiments took part, and emblematic figures of "Warfare and Courage" and "Grief." The composition is crowned with the winged figure of "Peace." In the small green near Park Place are two statues, one of Mr. John Cory, D.L., J.P., of Duffryn, and the other of Lord Ninian-Crichton Stuart, M.P., both by

Sir Goscombe John. Not far from it is one of the finest examples of the work of this eminent sculptor (himself a native of Cardiff), the equestrian statue of Godfrey, 1st Viscount Tredegar. In the public gardens facing the University College is a statue of Henry Austin Bruce, 1st Lord Aberdare.

The first statue erected in Cardiff was that of the second Marquess of Bute, by I. Evan Thomas. It now stands at the end of St. Mary Street, near the G.W.R. Station. One of Cardiff's earlier public men is commemorated by the statue of John Batchelor, "The Friend of Freedom," in the Hayes.

The Castle.

The present buildings of the Castle, the seat of the Marquess of Bute, are mainly a restoration by Burgess of the mediæval fortress. On the site have been discovered the remains of a Roman gateway and of a massive wall of Roman origin. Upon this foundation an earthwork fortification was erected at a later period. In the Middle Ages a wall enclosed the Keep and the mediæval Castle. The only portion of the Norman Castle remaining is the ruined keep, built by Robert, Earl of Gloucester, which stands on a moated mound in the centre of the court. The main entrance from Castle Street has a somewhat grim aspect, relieved by flower beds and the vines trained against the wall. At intervals on the outer walls are sculptured animals, the work of Mr. T. Nicholls. Next to the gateway is the Black Tower, which dates from the 13th century. It is linked to the Clock Tower by a massive curtain wall. This tower is modern, and formed part of an extensive scheme of restoration and addition made by the late Marquess of Bute. The interior of the Castle is lavishly decorated in the mediæval manner of which Burgess was such an interesting exponent. Burgess played an important part in the Gothic Revival of the 19th century, and students of that movement will be interested in the house he designed in Park Place, now used as the offices of the Llandaff and Dinas Powis Rural District Council.

St. John's Church.

With the exception of the Church of St. John, all the places of worship in the City are modern. This spacious Church, containing interesting memorials, was probably built in the last half of the 15th century. The tower, with its rich coronal of West Country type, is particularly fine, The best view of it is that obtained from St. Mary Street, looking down Church Street.

The Public Library.

On the other side of the Churchyard facing the Hayes is the Central Public Library. The original building was erected in 1882 to house the Library, Museum, and Science and Art Schools, but the three institutions soon outgrew it, and the Science and Art Schools now form part of the new Technical College in Cathays Park, and the Museum will be transferred to the new building of the National Museum.

An extension of the building, designed by Mr. Edward Seward, was opened in 1896 by King Edward VII., then Prince of Wales. The main front, crowned by a large figure of Minerva, faces the Hayes.

Theatres and Music Halls.

A little way up Queen Street, from St. John's Square, is the Cardiff "Empire," recently rebuilt and enlarged. Further on is Park Place, where the New Theatre, another recent building, is situated. The old Theatre Royal, renamed the Playhouse, in St. Mary Street, is at present in course of reconstruction from the designs of Messrs. Willmott and Smith.

Welsh National School of Medicine.

Just beyond the Taff and Rhymney Railway bridges, at the beginning of Newport Road, is the old University College, still used by the Faculty of Science, and the new building for the Welsh National School of Medicine, designed by the late Colonel E. M. Bruce Vaughan, and now in course of erection. The cost of this block of buildings is being borne by Sir William J. Thomas, Bart. Its completion has been delayed by the War.

Hospitals.

A little further up Newport Road is the King Edward VII. Hospital, formerly known as the Cardiff Infirmary. This hospital has been continuously added to and extended as the result of gifts by local benefactors, one of the most recent being the hospital church.

Not far away is the Prince of Wales Hospital for Limbless Sailors and Soldiers, which was established during the War. It consists of three houses with attached workshops, offices, and gardens on a site in the Walk.

There is also a Seamen's Hospital at the Docks.

The Fire Station.

In Westgate Street, opposite the Cardiff Arms football ground, is the Corporation Fire Station, another fine example of the work of Messrs. Harris and Moodie, the Architects of the Glamorgan County Hall. This building, with its lofty entrances boldly treated as a rusticated Doric colonnade, its field of soft-toned handmade brickwork relieved by stone only in the case of the first floor windows and the simple band which gives a frieze effect, its fine crowning cornice, Italian tiled roof and delightful hose-drying tower, is an excellent example of the sound architectural treatment of a utilitarian building.

Town Planning.

Students of Town Planning will be interested not only in the lay-out of the Civic Centre and the proposed widening of Duke Street, but in such features as the sports grounds in Westgate Street and the preservation of the banks of the River Taff which results in the possession by the City of a view from Canton Bridge which must be a source of great pleasure to many visitors and residents.

Llandaff Cathedral.

While not strictly in the scope of these notes, it may be mentioned that a short tram ride along Cathedral Road brings the visitor within a few minutes walk across Llandaff Fields to Llandaff Cathedral, a noteworthy mediæval structure ably restored by Mr. Prichard, a local architect. This building, in its peaceful setting, forms a striking contrast to the busy streets from which it can be reached so readily.

The date of the foundation of Llandaff is uncertain, but it is generally assumed to be the earliest of the Welsh Sees. Little is known of the early history of the Cathedral beyond the fact that the building of the present church was begun in 1120 during the time of Urban the first Norman Bishop. Its subsequent history was uneventful until the sixteenth century when it was despoiled, and from then until the nineteenth century it remained poor and in ruins. Its revival and rebuilding is mainly associated with the name of Alfred Ollivant, whose episcopate lasted from 1849 until 1883. Ollivant reawakened the religious life of the diocese, and saw the Cathedral completely and successfully restored.

The building itself is the only guide to the dimensions of the Church begun in 1120. It seems to have gradually extended westward. The Norman doorways in the third bay on either side are later than Urban's time and the early English Arcade later still. In the Decorated period the Norman building was partly reconstructed, but the fine Eastern arch opening into the Lady Chapel and part of the walls were left. The original screen is now in the north choir aisle.

In 1732, Wood was employed to build in accordance with the prevailing taste a Grecian temple extending from the high altar to the fourth bay, where a classical fount was erected.

Prichard swept this church clean away when he brought the Cathedral back to something like its earlier form, and reproduced so far as possible the original work. The North West Tower, built by Jaspar Tudor (uncle of Henry VII.), was given a new crown of pinnacles, but was otherwise unaltered. The other tower was too far decayed to be restored, and in its place a tower and spire of French character were built. The old doorway in the west front has survived the centuries of neglect.

The interior decoration of the restored Cathedral is of peculiar interest from the association of members of the Pre-Raffaelite School with it. Mr. Prichard left many of the details of the furniture to his partner Mr. John P. Seddon, who was intimate with the Pre-Raffaelite artists, and who sought their aid in the execution of his designs. He obtained models from Thomas Woolner for the four panels of the pulpit representing Moses with the stone tables of the Law; David harping; St. John the Baptist; and St. Paul, and the panels were executed from Woolner's models.

For the Sedilia in the Sacrarium a drawing was made of the Pelican in her Piety, by Dante Gabriel Rossetti, and used as the subject of one of the panels in the gables.

At Mr. Seddon's suggestion, the Cathedral Restoration Committee commissioned Rossetti to paint the triptych for the new Reredos, which Prichard designed in the Decorated style, with three gables. The subjects of the three panels of the triptych are David as Shepherd, the Nativity, and David as King. The figure of the Virgin was painted from Mrs. Morris, the wife of William Morris, and that of David Rex from Morris himself. Other figures are said to have been painted from Burne Jones and Swinburne. Some of the painted glass in the aisle windows and the frontal for the choir altar were the work of the well-known firm of Morris, Marshall and Company, with which William Morris was associated in his work for the revival of the decorative arts.

EDUCATION IN CARDIFF.

By HERBERT M. THOMPSON, Chairman of the Cardiff Education Committee and a Vice-President of the University College of South Wales and Monmouthshire.

PREFATORY NOTE.

In compiling the following record of forty-five years' work in building up the educational system of Cardiff, for the first half of the period I have relied largely on the work of my predecessors ; the late Charles T. Whitmell wrote the article on Education in the (Cardiff) British Association handbook of 1891, and Mr. Charles Morgan the one in the handbook of the Cardiff Exhibition of 1896. The Life of John Viriamu Jones, by his widow, is a mine of information for the period 1883-1901.*

During the latter half of the time I have had personal knowledge of what has passed from the circumstance that I have been a member of the principal educational governing bodies.

The official yearbooks, circulars, and other publications of the old Cardiff School Board, of the Cardiff Education Committee, and of the University College, give detailed particulars of most of the subjects treated, and from these I have had occasion to quote.

My thanks are due to Mr. J. J. Jackson, Director of Education, and Mr. D. J. A. Brown, Registrar of the University College, for having read the paper in manuscript and for giving me material help by their suggestions.

---

*e.g., It recounts the passing of the Welsh Intermediate Education Act and its results, and the establishment of the University of Wales (of which the University College of South Wales and Monmouthshire at Cardiff is a constituent). For neither of these subjects have I space in this already overcrowded paper but it should be noted that since there has been a University of Wales the courses of study in the Arts and in Science at the College have been primarily intended to be qualifying courses for the Degrees of the University in those faculties).

I have not considered it within the scope of this paper to speak of Schools under private management or Committee management within the City of Cardiff; nor have I spoken of those lying just outside the boundaries, though some of them, such as Howell's School, the large Secondary School for Girls at Llandaff, and the County Secondary Schools at Penarth, are important for the children of Cardiff citizens. If the project which is now on foot greatly to enlarge the City boundaries is carried through, many additional Schools will come under the Cardiff Authority.

As this is a handbook for visitors, I have given a little more space than might otherwise have been deemed sufficient to those features that show Cardiff as an *originator*, and distinguish its experience from that of most other large towns in the country.

I ought perhaps to mention that the facilities for the training of teachers include, besides the training departments at the University College, two residential Training Colleges in the near neighbourhood, one for women at Barry and one for men at Caerleon.

Before 1875.

The history of education in Cardiff has little to relate before the year 1875. It begins more abruptly than do most narratives of the kind, for in the middle of the last century first a flourishing town and then a great city arose where the community before had been of the dimensions of hardly more than a large village. The place, though of great antiquity, sprang into being as a centre of importance in a fashion more familiar to the New World or to Australia than to Britain, and like Colonial cities, it is almost without those endowed charities which are to be found in older places. One or two comparatively unimportant educational funds (conditioned by such customary picturesque details as blue bonnets for boys and distinguishing badges for girls) there were, but in the main, public education in Cardiff did not exist apart from the great national movements. But these, whether English or Welsh, left their mark on the local Schools.

At the beginning of the nineteenth century Joseph Lancaster had formulated his ideas and put them into practice in his School at Southwark. They were amongst the important influences which resulted in the establishment, in 1808, of the British and Foreign School Society, closely followed in 1811 by the National Society for Promoting the Education of the Poor in the Principles of the Established Church. "British" and "National" Schools sprang up all over the country, and Cardiff, like other places, was affected by these waves of educational enthusiasm; but such provision of School-places as was made did not keep pace with the rapidly increasing size of the town, and at the time of the passing of W. E. Forster's great Act in 1870, the attendance at Elementary Schools amounted to but one-third of what by present standards we should consider adequate for the population.

There was in the place no old-established endowed Grammar School like those of Cowbridge or of Swansea, but the Elementary Schools were supplemented by private ventures, many of them of the "Academy for Young Gentlemen" or "Seminary for Young Ladies" type, some, however, doing quite excellent and useful work, remembered even now with gratitude by not a few of the older ones amongst the citizens.

1875 (Population of Cardiff about 66,000).

The building of Schools.

The first great landmark date in Cardiff educational history may be said to be 1875, when a School Board was elected and began at once to build Schools, the first of which was opened in 1878. At the same time the Managers of "Voluntary Schools" (which by this time generally meant the National Society, representing the Church of England, or the representatives of some other religious body, such as the Roman Catholics or the Wesleyans) also began to build.

During the first twenty years of its existence the School Board built fourteen large "Board Schools," besides a Higher Grade School, whilst during the same period nineteen

"Voluntary Schools" (generally much smaller in size) were opened. Six of the latter were for Roman Catholics, the Irish population being large.

In most instances there were included in each of these 33 Schools three quite independent departments, each with its own head teacher, respectively accommodating boys, girls, and infants under seven years of age, so that in 1896 there were in effect nearly 100 Schools.

They were almost all in new buildings, and these were (for that period) of good type, the structures being airy and commodious, but their large class-rooms stereotyped what has been perhaps the worst blemish in our elementary educational system, viz., the classes of 60 or 70 pupils. Even at the present time in the majority of cases, the number in class is reduced little below 50. Each lustrum gave better facilities to the teachers for college training and witnessed a higher standard of work demanded by the public and attained by their own ambitions. The qualifications, therefore, of the staffs of men and women secured as teachers rapidly improved, but their *number* in each School was entirely insufficient. The Elementary Schools were at their inception, and continue to this day to be grossly understaffed, and so far as failure has been justly charged against the education they give (there is a large amount of success to put against it), it has mainly arisen from the mechanical methods, and absence of individual cultivation for the children, inseparable from such conditions.

Additions to School Curriculum.

The activities of the School Board were developed in many directions beyond that of the mere establishment of Schools. They, were made more complete by the establishment of Cookery Kitchens (1891) and Laundries for girls (1898), and of wood-work workshops for boys (1898-1901). One of the most interesting moments experienced by a visitor to a large Elementary School comes when he enters the spacious workshop, well equipped with its rows of carpenters' benches, and filled with boys evidently keenly interested in what they are doing ; yet it has been questioned, probably justly questioned, whether a less "intensive" course—one more generally distributed over the child's school-life, and more closely co-ordinated with other subjects of study, may not advantageously be adopted.

School Libraries.

In 1898 a serious attempt was made to induce the habit of home reading, and an interesting library system was introduced, which provided a good number of carefully chosen books for each Boys' or Girls' School Department in the town; the children of the upper standards (subsequently many of the younger ones also) could take home books for a week's reading. The care of these books and their issue were placed in the hands of the school teaching staff, but the whole scheme was under the control of the experienced authorities of the City's Central Library, who from the beginning have chosen the books and seen that the stocks are from time to time refreshed by a new assortment; once a year, too, they overhaul the whole of the books to see that there have been no considerable losses, and that the books requiring rebinding have attention. The system is maintained by a subsidy from the Education Authority equal to 6d. a year for each child (excluding infants) in School attendance, and has fully justified this rather large expenditure. The increased cost of books will now make a larger grant necessary.

1883 (Population of Cardiff about 92,000).

University College.

1883 was a great landmark date, for in that year little short of a revolution was brought about in the educational status of the town by the establishment of the University College of South Wales and Monmouthshire, under its brilliant young Principal, Viriamu Jones, and an equally youthful professional staff. The original staff consisted of nine Professors (the present number is 22) and five Lecturers and Demonstrators (the present number of Lecturers, Demonstrators, and Teachers is 51).

A group of earnest men had brought about this achievement, conspicuous amongst them being that great educationist, the first Lord Aberdare, Dean Vaughan of Llandaff, and Lewis Williams, Chairman of the School Board, and they reinforced the activities of the new Principal (Viriamu Jones) so that the remainder of the century was conspicuous for its educational progress.

The year after its foundation the College was incorporated by Royal Charter (1884), and gifts of £10,000 each were made by the Corporation of Cardiff and the third Marquess of Bute.

A temporary home was acquired in the premises that had recently been vacated by the Cardiff Infirmary, and the Faculties of Arts (with Chairs in Greek, Latin, Philosophy, English, and Celtic, and Lectureships in French, German, and Music) and of Science (with Chairs in Physics, Chemistry, Biology, and Mathematics) were established. The Government made an annual grant of £4,000 in support of maintenance.

Higher Education of Women.

It was a fundamental principle of the new College that all its privileges and duties, whether for students or staff, should be open and available for women on equal terms with men.

Hostels.

The presence of many women students at the College made it necessary to provide a Hostel for their accommodation, and one was opened in 1885 in premises which were adapted to the purpose in Richmond Road; it was called Aberdare Hall, after Lord and Lady Aberdare, who took a special interest in the project. In 1893 a fine new building was erected on the northern edge of Cathays Park, to which the students were transferred.* This building accommodates at present 60 students, but extension is urgently required, and at the moment subsidiary hostels have had to be brought

*This building is to be used as the "Secretarium" at the forthcoming meeting of the British Association.

into being to supplement the accommodation at Aberdare Hall itself. A second women's hostel has just been established, which will receive residents next term.

The need of a corresponding institution or institutions for men students has always been recognised, but has not yet been accomplished.

There is, however, a project on foot to establish a large students' Club for both men and women, somewhat corresponding to the Unions at Oxford and Cambridge, and this is the object to which the College War Memorial is to be devoted. £2,688 out of the £10,000 estimated as immediately required to establish the Union, has already been subscribed.

Higher Technical Department.

In 1890 the department of Applied Science and Technology was established, and Professors of Engineering and Mining were appointed. The Drapers' Company of London made a grant to the Engineering Department and founded a Scholarship. Soon afterwards the County Councils of Glamorgan, Monmouthshire, and Cardiff, made annual contributions to the Higher Technical Department and in consideration of this received free studentships at the College, the free students receiving also County grants for maintenance. Within recent years the Town Council of Newport has joined the scheme.

Training for Teachers.

Another departure of great importance in the same year (1890) was the establishment of a department for training teachers for Elementary Schools, 120 normal students being at once accepted.

Almost immediately afterwards provision was made for the training of teachers for Secondary Schools.

The Salesbury Library.

About 1887 a large collection of books and pamphlets about Wales, or by Welshmen, or in the Welsh language, was acquired for the College by subscription amongst some of its friends. It included a collection of engravings, many of them portraits beautifully executed in mezzotint. The Salesbury Library, as it is called, consists of nearly 20,000 volumes, and is this year, after too long an interval, being systematically arranged and catalogued so as to make it available.

When it is remembered that the Cardiff Public Library is probably the second largest important Welsh Library in existence (the first being the National Library at Aberystwyth) it will be seen that the Salesbury Library in conjunction with it, makes Cardiff a really important centre for Celtic study.

Secondary Education.

Whilst all this was going on at the College, the school Board established, in 1885, a Higher Grade School (now one of the Municipal Secondary Schools), consisting at that time of two parts, an Elementary School and, for the older pupils, an organised Science School.

Rather later (1889) the Welsh Intermediate Education Act was passed, and the small Committee appointed under that Act at Cardiff was not slow to establish under its provisions an Intermediate School for Girls (now the High School for Girls), followed a little later by one for Boys (now the High School for Boys).

It thus came about that whereas Municipal Secondary Schools sprang up about this time in the English large towns—and County Schools under the provisions of the Welsh Intermediate Education Act came into being all over Wales—Cardiff was in the almost singular position of being provided with Schools of *both* kinds (they number at present four of the first and two of the second type).

There was also established by the School Board a Pupil Teachers' Centre.

Evening Schools and Special Schools.

Evening Schools were also opened at first in 16 Schools (eight for boys and men and eight for girls and women) and Special Schools were provided (*a*) for the Deaf and Dumb, (*b*) for the Blind, and (*c*) (in 1901) for Mentally Deficient Children.

1901-4 (Population of Cardiff about 170,000).

With two important exceptions we. have now traced the principal local educational developments up to the beginning of the present century, and this forms a third natural landmark in our history, for in 1902, by the Education Act of that year the School Board, as well as the Governing Body of the Intermediate Schools and the Technical Instruction Committee, were to be merged in the new Local Education Authority, *i.e.*, a Sub-Committee of the Cardiff City Council—the actual transfer took place in 1904.

University College.

Meanwhile the University College was faced with the necessity of erecting new buildings. In 1895 a grant of £20,000 had been made by the Government to a Building Fund on the condition that at least an equal amount was raised from private sources, and the College collected the sum necessary to secure this grant. About the same time the (London) Drapers' Company undertook to contribute £10,000, but eventually erected as its own contribution the library at a considerably greater cost (£16,000).

In 1900 Cardiff Corporation made a free gift of five acres of land, of the estimated value of £20,000, for the new College building. The site was in Cathays Park ; the new College was therefore destined to be in the near neighbourhood of some five or six other great national or municipal structures, and thus to form an important part of a magnificent architectural scheme.

At this juncture Principal Viriamu Jones died (1901), aged only 45. He died just as the realisation of his dream of a worthy home for a great College came in sight. But though he did not live to see the actual structure, he had witnessed the coming into being of the departments in Science and in the Arts which were to fill it with life and beneficent energy. His successor, Dr. Ernest H. Griffiths, gave (before his retirement in 1919) an approximately equal period of years to carrying on this really great work, and it is difficult to overestimate what the intellectual development of the area owes to these two men.

I have said that our survey to the end of last century is incomplete in two particulars. Up to this point in our narrative, I have refrained from telling of the rise of Technical Education (with the exception of its highest branches, which, being of University standard, had already found a home in the University College) and of the Medical School, as the records of each appeared to demand uninterrupted narration.

TECHNICAL EDUCATION

So early as 1865, Technical Education had been provided in some small measure by the Science and Art Schools established by the Free Library Committee, and at that time necessarily supported out of the Library Rate (the only one available for the purpose). But in 1890 (after the passing of the Technical Education Act of 1889), the whole system was transferred to the Technical Instruction Committee of the City of Cardiff, and by it, for the time beng, placed under the care of the Council of the University College, who at once established four departments, viz., Pure and Applied Science, a School of Art, a Commercial Department, and a Women's Technical Department.

This departure quickly added more than 2,000 students to those who attended the evening courses at University College. The College Council recognised that the City of Cardiff, by thus entrusting its scheme of Technical Education to their care, was but emphasising the close and cordial relations that had always existed between the two corporations, and carried out their side of the contract so efficiently that the number of students grew rapidly, and with this growth took place an increasing pressure, eventually undue pressure, both on the College teaching staff and on the space at their disposal. Largely as a result of a report by Sir Philip Magnus in 1907, the Cardiff Education Committee realised that it must eventually make direct provision for this branch of its work. It accordingly appointed a Sub-Committee to supervise the work carried on by the College as well as the Evening Schools in various parts of the City. They appointed

Mr. Charles Coles to take charge of the department, and his energy and initiative called out the latent demand for a comprehensive system.

In 1908, after his report on the subject, a fresh grading was made of the subjects taught in the evening, and courses were introduced continuing over a period of years and, so far as is possible with evening classes, covering the whole field of Technical Education. These were arranged for Mechanical and Electrical Engineers, Fitters, Carpenters, Masons, Accountants, &c. At the same time the Continuation Schools held for less advanced pupils in the Council School buildings were turned into preparatory Technical Schools, and their curriculum was strengthened. There was also set up a *Day* Preparatory Technical and Commercial School for boys between the ages of 13 and 16.

Between 1908 and 1916, no less than nine separate buildings were brought into use in the evenings (besides the use of Council Schools for less advanced work), but since only two of these could be used by day, development in day work was hampered. Finally, in 1910, the City Council decided to erect new buildings, a free site of 1½ acres in Cathays Park, worth £6,000, was allotted by the City for the purpose. The new College was not actually opened till 1916. It is a well proportioned building, admirably suited to its purpose, built from the designs of a Cardiff firm of architects, who obtained the first place in an open national competition. It is interesting to note that the partners in this firm had themselves been pupils at the School.

Mr. Coles was now made Principal of the new College, and he soon found himself directing the studies of 500 full time and 3,500 part time students. The Technical College will soon be a completely post-Secondary Institution for students of 16 and upwards. To the departments ordinarily found in such Colleges in Higher and Intermediate Technological work in Commerce and in Art, there were last year added full time departments in Mechanical and Marine Engineering, Naval Architecture, and Industrial Chemistry, as well as in Architecture and Civic Design ; in addition, a Higher School of Commerce, a College of Pharmacy, a Marine Technical School, and a School of Bakery and Confectionery (the last opened early this year) were set up. Already an extension of the buildings is foreshadowed, and is indeed in some measure actually taking place.

THE SCHOOL OF MEDICINE.

We now pass to the School of Medicine. It was opened in 1894, when Chairs in Anatomy and Physiology were founded, and a new story was added for its accommodation to what were then the main buildings of University College in Newport Road. Tuition was provided for the first three years of medical study, the student proceeding to other Schools for study in Medicine, Surgery, and Gynæcology during the fourth and fifth years. The Welsh students therefore were obliged to take their degrees or diplomas at Universities or at Institutions not in Wales. Nevertheless, within the limits of its responsibility, the results attained by the College were brilliant.

In 1899 post-graduate tuition for the diploma of Public Health was provided.

In 1906, the University of Wales obtained power to confer degrees in Medicine and Surgery, a power, which now at last, when the Medical School at Cardiff is about to become complete for the full course, is beginning to be exercised.

In 1909 a new Treasury grant made possible the establishment of the Department of Pathology.

Advance was now quickened by a series of munificent gifts to the School. From 1912 to 1914, Sir William James Thomas made a continually progressing series of promises of aid. They have already resulted in the erection of the Physiological Block (designed by the late Col. Bruce Vaughan, a Member of the College Council, who was throughout closely identified with the new movement), and the £30,000 or more so expended is being followed by a sum twice as large for the necessary buildings to complete the School so far as the professional subjects are concerned, including a Public Health Department and a School of Preventive Medicine.

Meanwhile a gift of £37,500 Railway Debenture Stock from the late Miss Talbot, of Margam, has secured a stipend of £1,500 a year for a Professor of Preventive Medicine.

The departments of Pure Science in which medical students must pass their first year, and other scientific students a longer period, are still wretchedly housed. The five great laboratories necessary to give the Pure Sciences a worthy home are to be built on the Cathays Park site, and Lord Glanely, the present President of the College, has given £25,000 for the erection of one of them, viz., the Chemical Laboratory.

1920 (Population of Cardiff about 213,000).

In concluding this paper, I may be allowed to glance for a moment or two at the developments immediately in prospect if our present ideals are to be realised.

University College.

Speaking first of the University College, the building scheme on the original site in Cathays Park ought almost immediately to be brought nearer completion by the erection of the five Laboratories and Departments of Pure Science that have already been spoken of, and if in addition there were built the Great Hall contemplated in the original plans, this part of the site would be almost occupied.

The Cardiff City Council being convinced that further development would be impossible without enlargement of the site has this year granted an extension of about 3¼ acres. This most timely gift will enable the Department of Agriculture, which Lord Glanely has promised to erect, to be proceeded with, and will give some room for the expansion in other directions that the recent increase in the number of students has rendered inevitable.

The old College site in Newport Road, greatly enlarged by the aquisition of freeholds in the vicinity will, in the main, be devoted to the new Welsh Medical School, although the Enigineering Department and that of Metallurgy will also find their homes there. With regard to the Medical School, it is almost inevitable that it shall be recognised as a constituent part of the University College of South Wales and Monmouthshire, for although the Haldane Royal Commission of 1918 reported in favour of constituting it as a separate College of the Welsh University, the sentiment, not only of those in Cardiff, but throughout Wales, as expressed through practically all the representative bodies concerned in the matter, is unanimously against severance, although it is clear that special arrangements for the government of the Medical School within the College must be made. This view has the concurrence of the Authorities of King Edward VII.'s Hospital, itself a very important, even it may be said, an essential partner in the enterprise.

The financial side of the development of Education of University type will shortly it is hoped be immensely helped (not only in Cardiff but throughout Wales) by the great scheme by which it has been determined that each County and County Borough in Wales shall levy a penny rate, and thus raise a sum which, supplemented by an equal amount from the Tresaury, shall foster University Education in Wales.

Space does not allow any detailed account of the Haldane Commission referred to above. It must suffice to say that its recommendations involve far reaching developments and much reconstruction in the College.

Mention should be made of a scheme for the regulation of Technological Education and Research. This scheme is based upon the co-operation of the University Colleges at Cardiff and Swansea, with Technical Colleges and Schools, and with a "Business Committee" consisting of representatives of the leading industries of South Wales and Monmouthshire. For the purpose of carrying the scheme into effect it is proposed that a Faculty of Technology and a Board of Technology should be constituted. These bodies would possess wide powers, but in certain definite ways they would come within the organisation of the University and the Colleges. It is an interesting proposal, and so far as I know one that has not a parallel in other parts of the country.

It would lead to such co-ordination that it would be possible for the different Colleges to work in co-operation, the specialities of each being made available for the rest, thus avoiding a cut-throat competition under which each institution would be inclined to attempt everything, but might accomplish nothing supremely well.

As a means, too, of securing improvement in industrial processes and methods, the proposed scheme is bound to have a far-reaching effect.

Carrying out the New Education Act.

Turning to that part of the Education system which is governed by the Cardiff Education Committee, it is inevitable that the new Education Act will bring about most important changes within the next two or three years. Under that Act we are directed to provide education at Central Schools (or Schools of a similar type) for children whose parents may desire it up to the age of 16, and for other children Day Continuation Schools, which shall provide the equivalent of something like one complete school day for 40 weeks in the year ; the pupils in these will also, for the present, be young people up to the age of 16, though ultimately the age will be extended to 18.

The Cardiff Scheme is to create centres serving large districts of the town. Sites will be acquired, on each of which a School of the Central type will be built, and in close proximity another School for Day Continuation work. In both types of School the size of the classes is not to exceed 30 pupils. It is hoped to secure good playing fields in the near neighbourhood of the Schools. Of the new School sites there will at first be three or four, but the number will probably have to be added to before many years have passed. The actual scheme for their government has not been quite decided upon, but there will probably be an intimate link between the two schools on each site.

Turning from the compulsory to the permissive clauses in the Act, it is hoped to establish an independent nursery school in one of the poorest districts of the town, as well as a smaller one, also of the character specified in the Act, to be comprised within one of the present Infants' Schools. An Open-Air School for ailing children is also contemplated.

Twelve to Fourteen.

In Cardiff, as elsewhere, there is no part of the present day education so weak as that given in the Elementary Schools between the ages of 12 and 14. Its inadequate character is the cause of great dissatisfaction to parents and children alike, and accounts for the far too punctual observance of the privileges obtained on attaining the legal leaving age. The reason is principally that, except in the larger schools, the number of pupils is not sufficient to have separate classes for the 6th, 7th, and ex-7th Standards. The consequent lack of classification means that some children are merely repeating what they have already been through (perhaps more than once)—"marking time" as it is ironically called; on the other hand, some are put to work too difficult for them, and so become discouraged. The most vital conditions of success in obtaining greater freedom for fostering individual bent, instilling more vital knowledge of the facts of mature life, and encouraging greater responsibility of self-determination in the pupils themselves, are placing children together to form separate classes in each stage of attainment, and securing that these classes shall be moderate in size.

The Cardiff Committee proposes to meet these difficulties by providing that children shall enter the Central Schools, not at the age of 14, but at 12.

With regard to children who in all probability will leave School at 14, and will therefore not enter the Central Schools, the proposal is to establish classes of superior type (no class to contain more than 40 pupils) for each of the Upper Standards, in each district in some School or Schools where children of this age can be concentrated ; they would not then be left in Schools where their number would be so small as to prevent efficient organisation.

New Conditions for School Children.

In recent years, Education Authorities have assumed responsibility for aspects of child life more and more removed from what occurs in the schoolroom. On the one hand an elaborate system of medical inspection and remedial health centres has grown up. On the other hand the leisure of the children, especially of the more neglected amongst them, has been the subject of concern for welfare committees which in every district are busily running clubs and keeping in touch with children and their parents. At the same time, and working closely in touch with the latter, Juvenile Employment Committees have sought to make the transfer from school to working life one that will establish the children in avocations suited to their character and abilities.

In Cardiff all this has been going forward and systems of considerable elaboration for health-care, for welfare work, and for assistance to the best possible employment have grown up. This City was, for example, one of the earliest to adopt the Choice of Employment Act and to appoint a Juvenile Employment Officer. It now possesses six after-care Committees with Secretaries, each of whom is paid a small stipend and under the fostering care of these Committees, two clubs for boys have come into being. Besides the ordinary Medical School inspections, clinics for minor ailments have been established (aural, optical, and dental). The medical work is closely linked up with that for baby welfare undertaken by the Health Committee.

Considerable as these developments have been they are dwarfed by the proposals of the new Act. That every working lad and girl must give up the equivalent of one working day a week to attend school till the age of 16 (later on till the age of 18) entails something like a revolution in the conditions of social and economic life, and will obviously require the utmost tact, forbearance, and goodwill on the part alike of Education Authorities, Employers, Parents, and Children. Difficulties smaller in degree but of the same character will accompany the drastic restriction of the employment of children out of School hours before they are 14.

Yet it is obvious that if all these far-reaching proposals can be successfully carried through, our School Teachers will soon have in their classes pupils whose home and occupational conditions, as well as their health, will produce results that ought far to outstrip what has been possible in the past.

The survey of the last forty-five years that has been attempted in this paper shows that an edifice of great extent and wonderful elaboration has been evolved almost out of nothing; but throughout the narrative, and especially in its last section, will be found indications that the tasks that must be accomplished during the next forty-five years are no less exacting.

THE CARDIFF PUBLIC LIBRARY.

By HARRY FARR, F.L.A., City Librarian.

THE Council of the Borough of Cardiff was the earliest authority in Wales to adopt the Public Libraries Act of 1855. After an unsuccessful attempt to adopt the Act in 1860, a voluntary library was started in 1861; but in 1862 another attempt was made, which succeeded. Cardiff, therefore, has had its public library for 58 years. It developed with the rapid growth of the town. Housed originally in a single room in the Royal Arcade, it was transferred in 1864 to a building belonging to the Y.M.C.A. in St. Mary Street. Eventually a new building was erected and opened in 1882, which included, in addition to the Public Library, a Museum and Science and Art Schools. These institutions soon outgrew the building. The Science and Art Schools were transferred to a newly formed Technical Instruction Committee in 1890. The Museum was transferred to a separate Committee in 1893, and later was handed over to the National Museum of Wales. A portion of the building is still occupied by the Museum until the new buildings in Cathays Park are ready to receive the collections not yet removed from the Library.

The needs of the Library, however, could not wait for these developments, and an extension of the building was begun in 1893, and completed and opened by King Edward VII. in 1896. This comprised the portion of the building extending from the Museum entrance in Trinity Street to the Hayes. In the new portion of the building are the General Reading Room, the Women's Reading Room, and the Reference Library. The Lending Library occupies the ground floor of the old building, and the Museum the first and second floors.

The Library and Science.

The Central Library contains large collections of books and periodicals on the pure and applied sciences. In the Reference Library will be found large scientific collections, together with the Transactions and Proceedings of a large number of scientific and learned societies and other scientific and technical periodicals, which are received regularly. These are either subscribed for or presented and are filed and bound for permanent reference. Official and technical publications, such as the Collected Researches of the National Physical Laboratory and the publications of the Engineering Standards Committee are also received regularly. The publications of the Patent Office are supplied as issued, and the Library has a complete set of these publications from the beginning.

The Scientific sections of the Library are continually being added to, and important monographs on special subjects are purchased from time to time. The following selected list of special monographs in two important branches of Natural History will give some notion of the Reference Library collections on special subjects :—

Botany.

Duke's *Genus Iris.*
Elwes and Henry's *Trees of Great Britain and Ireland.*
Engler and Prantl's *Die naturlichen Pflanzenfamilien.*
Hooker and Jackson's *Index Kewensis.*
Millais' *Rhododendrons.*
Oudemans' *Enumeratio Systematica Fungorum.*
Ravenscroft's *Pinetum Britannicum.*
Sanders' *Reichenbachia.*
Sowerby's *English Botany.*

Zoology.

Bronn's *Klassen und Ordnungen des Tier-Reichs.*
The Fauna Hawaiiensis.
Barrett's *Lepidoptera of Great Britain.*
Wytsman's *Genera Insectorum.*
Reeves' *Conchologica Iconica.*

Buller's *Birds of New Zealand.*
Gould's *Birds of Great Britain.*
Birds of Europe.
Birds of the Himalayan Mountains.
Trochilidæ.
Sharpe and Wyatt's *Swallows.*
Thorburn's *Birds of Great Britain.*
Elliott's *Felidæ.*
Millais' *Mammals of Great Britain.*

A noteworthy gift to the scientific collections in the Library was made by Mr. (now Alderman) H. M. Thompson in 1891, when he purchased and presented a portion of the library of the late Professor W. Kitchen Parker, F.R.S. This valuable gift included a complete set of the Reports of the Challenger Expedition in 50 volumes, and also sets of the Transactions and Proceedings of the Royal Society, the Ray Society, the Linnean Society, and the Zoological Society.

Important gifts have also been received from time to time from the British Museum (Natural History), including the Scientific Reports of the National Antarctic Expedition, 1901-4; from the U.S. National Museum; and other sources.

The Library has a number of early books on Science, and these have been arranged and catalogued for exhibition during the visit of the British Association.

Provision of scientific and technical literature is, of course. only a part of the work of the Library, and some account of the general work and activities of the Library may be of interest.

THE REFERENCE LIBRARY.

In their second annual report, dated 9th November, 1864, the Library Committee remark: "The Committee . . . desire to create a Standard Reference Library of those books which are so expensive as to be found in few private libraries." This Reference Library, begun with a nucleus of 49 volumes, has been continuously added to. Not only have important books and collections been constantly purchased by the Committee, but a long succession of generous benefactors have presented valuable collections, thus helping to build up a Reference Library which now numbers some 120,000 volumes, a stock which is only exceeded by the municipal Reference Libraries at Manchester, Liverpool, Birmingham, and Glasgow.

The reference collection which the Committee keeps before them as their ultimate ideal may be summed up in the words "to every student his book." This ideal collection is far from having been reached, but the comprehensive collection of books on all subjects in the Reference Library will meet the needs of everyone except the very special student, and it is very rarely that a request for books or information cannot be satisfactorily met.

Special Collections.

In addition to the main reference collections, the Reference Library includes a number of special collections.

Celtic Collection.

The most important of these is the Welsh and Celtic Collection. Welsh newspapers and periodicals have been taken ever since the Library was established. Gifts and purchases of books in Welsh and relating to Wales are recorded continuously in the early reports. These, though at first few in number, gradually grew more numerous. Later, the Committee definitely adopted the policy of acquiring systematically books in Welsh and in the allied Celtic languages and of books on the history, topography, antiquities, language and literature of Wales and the other Celtic countries. This collection now contains some 50,000 volumes and pamphlets, its only real rival being the National Library of Wales. It is regularly used by Welsh scholars and students and others interested in Celtic subjects from all parts of the Kingdom, and occasionally by students engaged in special research work from continental and other countries.

Manuscripts.

The MSS. proper number over 800, of which 110 have been calendared by the Historical MSS. Commission. They include a large number of MSS. in Welsh. The earliest dates from the 13th century, and is a famous MS. known as the "Book of Aneirin," and there are other Welsh MSS. of the 13th, 14th, 15th, 16th, 17th, and 18th centuries. The collection also includes heraldic MSS., pedigrees and pedigree rolls, court rolls, and other historical and genealogical MSS. of considerable importance.

There is also a small collection of illuminated and other MSS. dating from the 13th to the 17th centuries, including Books of Hours beautifully illustrated and decorated with miniature paintings. It is hoped to add to this small collection from time to time, and make it of real service to the student of palaeography and mediaeval art.

In addition, the MSS. include a large collection of separate deeds and documents mainly relating to Wales, numbering over 1,500 items. The earlier deeds are mostly in Latin, and date from the 12th century.

Prints, Drawings, and Photographs.

The Library has a very large number of prints, drawings, and photographs, mainly topographical. The most important item is the collection of water colour and other drawings by Charles Norris, chiefly of Pembrokeshire subjects.

Other Special Collections.

There are a number of other special collections nearly all of which have originated in gifts made from time to time. Amongst them are a collection of books printed before the year 1500 (incunabula) from the early printing presses of Germany, Italy, Switzerland, France, and the Low Countries: early editions of the Classics; a collection of English books printed before the year 1640 and later 17th century English literature; an English drama collection; 15th, 16th, and 17th century Italian, French, Spanish and Dutch books;

emblem books and other early illustrated books; books illustrated by Cruikshank, and other 19th century English illustrated books; books printed at the Kelmscott, Doves, and other modern presses; a collection of early children's books and early educational literature; contemporary French Revolutionary literature; a collection of 17th and 18th century music, which it is hoped to develop into a comprehensive music reference collection; a collection of Quaker literature; and other smaller collections.

The Central Lending Library.

The Central Lending Library is a wide and comprehensive collection of books in all branches of literature, and is intended to meet any reasonable demands that may be made upon it. The standard books, text-books, treatises, and popular books on all subjects are well represented on its shelves. The books are grouped and classified, the main divisions being General Works and General Science; Mathematical, Physical, and Engineering Sciences; Biological and Medical Sciences; Anthropological and Occult Sciences; Social and Political Sciences; Philosophical and Theological Sciences; Art and Archaeology; Language and Literature; History, Biography, and Topography. Included in the Arts Section is a large and representative collection of musical works, and in the Language and Literature Section a representative collection of texts of the works of modern European writers, including French, Italian, Spanish, Portuguese, German, Flemish, Scandinavian, and Russian. The total stock of the Central Lending Library is over 45,000 volumes.

Branch Libraries.

As the population of Cardiff grew, and the town gradually extended farther into the adjoining suburbs, demands began to be made for branch libraries and reading rooms in the outlying districts. In 1898 the Corporation secured Parliamentary powers to levy a 1½d. rate for library purposes, and a scheme for the provision of branch libraries was adopted. There are now six branch libraries, in the Cathays, Roath, Canton, Grangetown, Docks, and Splotlands districts.

School Libraries.

In addition to the branch library service, a School library service was organised in 1899, jointly by the Library Committee and the Cardiff School Board. Each Elementary School department, and also the Secondary and Pupil Teachers' Schools, now have school libraries for the use of their scholars. The total stock numbers over 25,000 volumes, and the annual circulation exceeds 300,000 volumes.

Special Services.

A number of special services have been undertaken in order to make the libraries as useful as possible to all classes of the community.

Telephone Inquiry Service.

A special feature of the library service is the answering of enuqiries by telephone. This was started in 1907, and has steadily increased in usefulness to business men and the public generally. Some 3,000 enquiries are dealt with in this way every year. They include enquiries for telegraphic and other addresses, the meaning of cable code words, and other information required daily in commercial and other offices. Enquiries for special information on a wide range of subjects are also continually being made and answered.

Loans of Illustrations.

Frequent requests having been received from teachers and others for the loan of illustrations in order to illustrate lessons, lectures, etc., a number of mounted and other illustrations were got together and arranged in groups for this purpose. These have been added to continuously and are largely used.

Holiday Service.

During the holiday season a collection of local guides, and other literature relating to holiday resorts, is displayed in the Reference Library for the use of people arranging their holidays. Special arrangements are also made by which books can be borrowed for extended periods for holiday reading.

Future Developments.

The growth of the library has been continuous since it was first established, and it will without doubt continue to grow and develop. Any scheme for the extension of the boundaries of the City, if carried through, would require a reconsideration of the present system of libraries. The Committee have also under consideration the new situation which has developed owing to the war and the increased demands which are likely to be made on the libraries as a result of the new social and educational movements. If greater interest is taken in scientific and technical research, and greater demands are made on the library in consequence, every effort will be made to meet them.

Use of Libraries.

Anyone who lives, works, or pays rates in Cardiff is entitled to use the lending libraries. Non-residents may also use the lending libraries on payment of 2s. 6d. a year. The Reference Library and Reading Rooms are freely open to all comers, subject to the library rules and regulations.

Every facility will be given to members of the British Association who wish to use the Library during their visit.

THE HISTORY OF MUSEUMS IN CARDIFF.

I. The Museum as a Municipal Institution.

By JOHN WARD, M.A., F.S.A., Late Curator of the Cardiff Museum; Late Keeper of the Archaeological Department of the National Museum of Wales.

The early history of the Museum is closely entwined with that of the Free Library, for both were departments of one institution, the Cardiff Free Library and Museum, and remained so until 1893. The earliest intimation of a movement for a free library for Cardiff was some correspondence in the local papers in 1858, in which the late Mr. Peter Price took a prominent part. Although nothing further of a public nature was done for two years, the movement gained ground, and this was largely due to Mr. Price's energy and enthusiasm. In October, 1860, a public meeting was held to consider the adoption of the Public Libraries Act, and, although poorly attended, a considerable majority of those present were in favour; but several days later, at a meeting of burgesses at the Town Hall, there was a majority of one against the adoption. Nothing daunted, however, the promoters thereupon opened a subscription list for providing a voluntary free library, and £60 were collected in the room. A fortnight later, they elected a committee, treasurer, and honorary secretary, also a number of gentlemen as vice-presidents, with Colonel Crichton Stuart, M.P. for the borough, as patron. Early in January, 1861, they issued a circular setting forth their aims in the establishment of a "Cardiff Museum and Free Library," and soliciting subscriptions (which now amounted to nearly £150) and "donations of books and objects suitable for the Museum,"—"the property thus acquired to be vested in the Corporation as a free gift to the town for ever." It is clear that a museum was contemplated from the first, not only from the name of the proposed institution, but also the provision of a room in it "for the reception of Antiquities, Curiosities, Botanical and Geological specimens and objects in Natural History." Several months later (on June 15th, 1861), a room at the St. Mary Street end of the Royal Arcade was opened as a news room, and a librarian was appointed at a salary of £20 a year, out of which he had to pay for a boy to assist him! Presumably the room was opened for only a few hours each day, and the small stipend supplemented an income derived from some other employment. Subsequently an adjoining room was rented for a lending and reference library. The experiment was a success, and in November the Libraries Act was adopted and the small institution became the property of the town. The corporation grant was £450 a year, and this continued until the adoption of the full penny rate in 1867. The librarian's salary was raised to £60, out of which he had to defray the cost of cleaning and firing, but not the boy's wages.

It was soon obvious that more space was needed, and this led to the premises of the Young Men's Christian Association on the opposite side of St. Mary Street being taken on a lease at £100 a year, from January 1st, 1864. A Museum Sub-Committee was appointed; "a back room in the top storey" was set apart for museum purposes, and the purchase of glass cases and fittings at a cost not exceeding £50 was sanctioned. Cards soliciting the gift of "specimens illustrating the local Natural History, Botany, Mineralogy and Antiquities of the County of Glamorgan," were printed and distributed. Gifts followed, but tardily. The most important of the earliest acquisitions was a considerable number of fossils presented by Mr. Price, which linked the new museum with two previous Cardiff institutions. The earlier of these was the Cardiff Athenæum of which Dean Conybeare, the great geologist, was the moving spirit. This institution fell into difficulties and its collection, or most of it, was removed to the Cardiff Library and Scientific Institution. In its turn, this fell on evil days, and about 1856, the trustees of the late Marquis of Bute, the owner of the premises, removed the furniture and other effects, except most of the geological specimens, and as the premises were shortly afterwards leased to Mr. Price, these passed into his hands. Eventually, in 1869, Lord Bute transferred the library and the residue of the collection to the Free Library and Museum.

In 1866, the Committee's responsibilities were increased by the establishment of Schools of Science and Art on the premises. Small wonder that in their report for this year the Committee regretted that in consequence of the additional expense thus incurred, no purchases could be made for the Museum! The year 1867 was eventful in the annals of the Museum. An energetic, if somewhat erratic, young man, Mr. Philip Stuart Robinson (better known in after years as "Phil Robinson," the author of several popular works on natural history and other subjects) was now Librarian. The "back room in the top storey" still remained a storeroom and closed to the public; but he determined to put an end to this state of affairs, and with this in view, induced several local naturalists to support him in an application to the sub-Committee for glass jars and other natural history appliances. The sum of £2 10s. was placed at their disposal for the purpose. Out of this incident came the Cardiff Naturalists' Society, one of the chief aims of which was the development of the Museum, and Robinson was the first honorary secretary. The new society started well, and the result was a greatly increased influx of gifts to the Museum. Robinson left Cardiff in 1868, but the Museum found in the President of the new society, Mr. William Adams, C.E., F.G.S., not only a generous donor, but a resolute organizer. The whole of the top storey was handed over to the Museum, and was converted into a single large room by the removal of partitions; and Prof. R. Etheridge, F.R.S., was engaged to name and arrange the palaeontological collection. At length, after the vicissitudes of ten years, the Museum was opened to the public on Wednesday and Saturday evenings, and on those occasions was in charge of a library attendant.

The first curator was an honorary one, Mr. John Williams, proprietor of the "Glove and Shears," who undertook in 1874 to take charge of the natural history collection. Two years later the coins and antiquities were similarly placed under the Rev. W. E. Winks' care. Two honorary curators continued to be appointed by the Naturalists till 1892. During the three years, 1874-6, the Museum was indeed a starved institution, for nothing was specially spent on it. In the following year, the parent Committee was asked "to allocate £100 a year on the Museum for a curator and other purposes." This, so far, had the desired effect, that a few months later, John Storrie, who will come again before the reader, was appointed Curator, at 9d. an hour, the Museum to be open 30 hours a week; he, however, resigned in the following year, 1879. With a view to a successor, it was ordered that "an attendant having some scientific knowledge be advertised for at a salary of £1 a week in *Nature*, *Science Gossip*, and local papers." This remarkable advertisement probably did not appear, for within a week Mr. M. H. Cochrane, F.C.S., was engaged on the same terms as Mr. Storrie. After thirteen months he resigned, and he was succeeded by Mr. A. C. Crutwell, F.G.S., F.R.H.S., who had already given lectures on geological subjects in the Museum. During his short spell of office—he resigned in less than a year—he named some hundreds of fossils and completed a manuscript catalogue of the collections begun by his predecessor,—all on a salary of £60 a year! By this time, the authorities realized that they must offer a higher salary, and in June, Mr. Storrie was re-engaged at £100 a year, and he held the post till the end of 1892.

Meanwhile as early as 1872, it was evident that more commodious premises would have to be provided for the growing institution, and for some years there was "a battle of the sites." The final selection was the site covered by the older portion of the Central Library, Trinity Street, and still occupied by the Museum. The new building was opened with great éclat on May 31st, 1882, and the day was observed as a public holiday. The event was a turning-point in the history of the Museum. Hitherto it had been almost exclusively devoted to natural history; but the gift of a large painting (one of Vicat Coles' masterpieces) by Sir Edward Reed, M.P. for Cardiff, and six others, with twenty-two examples of Nantgarw and Swansea porcelain by the Committee of a local exhibition in 1881, and the bequest by Mr. Menelaus of Dowlais of thirty-nine paintings, widened the scope of the Committee's ambitions.

One would have expected that in providing a new home, foresight would have been exercised to ensure its sufficiency for a generation. Within three years the Committee reported that "there is already pressure in some departments of the Institution and that it is certain that an extension will be necessary at no distant date." Cardiff, however, was now growing with a rapidity which probably no one had anticipated. During the decade beginning with 1881, the population increased by more than 50 *per cent.*, the rateable value more than doubled, and nearly 11,000 new houses and shops were erected. The subsequent annual reports were more and more insistent upon not only the need of extension, but of increased funds. By 1889, it was evident that the proposed extension would be inadequate, and this led to the transference of the Schools to a new authority with a view to their early removal; and at the close of 1892, the Museums and Gymnasiums Act was adopted. The Schools were removed in 1892 and the vacated rooms were divided between the Library and the Museum; but through a number of unforeseen circumstances, in which shortness of capital was prominent, the latter is still where it was—in Trinity Street.

With the adoption of the above act, the Museum lost a remarkable personality in Mr. Storrie's resignation. He was a Scotchman who, about 1872, found employment in the *Western Mail* printing works, and whose botanical and geological knowledge soon brought him into contact with the Naturalists' Society. His versatility was remarkable. There seemed hardly a subject that he had not dipped into at some time or other. This, together with his rough-hewn exterior, northern brogue, and careless attire, conduced to make him a notable and popular character, and in his declining years he was often referred to as the "Cardiff Sage" in the local papers. A keen collector, he added greatly to the natural history section of the Museum, but his lack of business qualities and hasty temper made him less successful as a curator. He was the author of a "Flora of Cardiff and East Glamorgan," and a local Silurian alga, *Nematophycus storrei*, was named after him as its discoverer. The present writer became his successor in May, 1893.

With the adoption of the Museums and Gymnasiums Act, the Museum made rapid headway. From the first, it had been a starved institution; and although after the removal to Trinity Street, the annual sums spent on it were considerably increased, they only averaged £330 a year, whereas under the above Act, they gradually increased from £1,733 in 1894 to £2,120 in 1912, and the expenditure on the purchase of specimens in 1906-8 was ten times that in 1886-8! The main policy of the new authority was the early provision of a larger home and the preparation of the collections so as to be ready for removal thereto, and when it became evident that Welshmen desired a National Museum, the movement had its energetic support, as also the claims of Cardiff to be its seat.

The first step towards improvement was to get rid of the old collection of birds, with the exception of the rarities, as most of the specimens were in a faded or dilapidated condition, and to initiate a new collection, confined to Wales, and mounted in standardized cases. The increased space consequent upon the removal of the Schools, admitted of a better distribution of the exhibits, which was carried out in 1894-5. To the two recently-acquired lower rooms were removed the antiquities, porcelain, topographical prints, and the like, and this allowed of the two smaller rooms upstairs being treated as art galleries, natural history being as before confined to the large room on that floor. Meanwhile, the important step was taken of forming a collection of casts of the Pre-Norman sculptured and inscribed stones of Wales, a truly national work, which would have been practically completed by now, had the recent war not intervened. The six years following 1897, were notable for the acquisition of Nantgarw and Swansea porcelain, with the result that this collection became one of the best in existence. In 1898, the art department was enriched by the gift of 151 oil-paintings, water-colours, and etchings, by the executors of the late Mr. J. Pyke Thompson, and in 1902, Lady Maud Vivian presented her late husband's cabinet of British Macro-Lepidoptera. In the same year, with the view of developing the zoological collection and preparing it for a new museum, a zoologist was engaged, and a large room was rented, as his workroom; and co-incident with this, was begun a collection to illustrate the art of the old English potter, which has been prosecuted with great success. In 1893, Welsh pre-history was illustrated in the Museum by only about half-a-dozen objects, and progress in this respect was slow till 1905, when a hoard of Late-Celtic bronzes from Seven Sisters, near Neath—the most important Welsh find of the kind—was presented. Two years later was acquired the Lloyd collection of the 17th-century tokens of Wales and Monmouthshire. For some years prior to this special attention had been given to the acquisition of objects illustrating old-fashioned life, especially in Wales, and in 1906 this collection of "by-gones" was considerably enlarged by the purchase of old lighting and other appliances at the Bidwell Sale. The Museum possesses a good collection of the old japanned goods of Pontypool and Usk, most of which were purchased from descendants of the last proprietors of the factories in 1903 and 1911. The last important acquisition of the Corporation Committee was the huge collection of pre-historic stone implements made by the late Mr. Stopes.

In 1901, the name of the institution was altered to "The Welsh Museum of Natural History, Arts and Antiquities," as better expressive of "the growing national character of its collections." In 1905, events were moving rapidly towards the realization of a Welsh National Museum.

Early in the year, H.M. Treasury appointed a special Committee, consisting of Lord Balfour of Burleigh, K.T., the Earl of Jersey and Lord Justice Cozens-Hardy, to determine its location. With a view to strengthen the claim of Cardiff for the Museum, the Museum Committee passed a resolution urging the Corporation "to offer the Committee of the Privy Council . . . . (*a*) the whole of the collections in the Welsh Museum of Natural History, Arts and Antiquities; (*b*) the site in the Cathays Park allocated for a new museum; (*c*) a capital sum of £7,500; (*d*) the capital sum of £3,000, being the gift in memory of the late James Pyke Thompson, J.P., and the sum of £3,000 further promised by his executors conditionally; and (*e*) the annual amount produced by a rate of one halfpenny in the £." In the following June, the Treasury Committee recommended Cardiff for the National Museum, and on November 15th, 1912, the union with the National Museum was completed by the transfer of the Corporation collections and staff.

II. The Museum as a National Institution.

By A. H. LEE, M.C., Secretary to the Museum.

Whilst the foregoing pages give some account of the work which was done locally with a view to providing Wales with a National Museum, it may be interesting to record what steps were being taken by Welsh Leaders in the Houses of Parliament. It is nearly twenty-five years since a few enthusiasts raised the question in the House of Commons. Among these pioneers may be mentioned Sir Alfred Thomas (now Lord Pontypridd), the late Mr. William Jones, Colonel Pryce-Jones, The Right Hon. J. Herbert Lewis, and the late Mr. Tom Ellis. After long discussion the matter was brought to a head on March 10th, 1903, when a resolution approving of the scheme was moved in the House of Commons by the late Mr. William Jones, seconded by Mr. J. Herbert Lewis, supported by Sir Alfred Thomas and others.

On the 19th March, 1907, His Majesty King Edward VII. was pleased to found and incorporate the National Museum of Wales by Royal Charter defining the objects of the Institution, and its government by a Court of Governors, Council, and Officers therein named and appointed.

The Council, having taken the best advice available in order to arrive at an accurate estimate of the requirements for a suitable building, invited an open competition for plans. One hundred and thirty designs were submitted, and the assessors, Sir Aston Webb, R.A., Sir J. J. Burnet, A.R.S.A., and Mr. Edwin T. Hall, F.R.I.B.A., awarded the first place to the designs of Messrs. Smith & Brewer, of Queen Square, London, to whom the work has accordingly been entrusted. Building operations commenced in 1910, and on the 6th November, 1911, His Majesty King George V. was pleased to grant a Supplemental Charter extending the powers granted by the original Charter, and was further graciously pleased to signify his special interest in the Museum by laying the Foundation Stone of the building at Cardiff, on the 26th June, 1912. Work continued steadily until 1915, when the superstructure of that part of the building which it was decided to proceed with was in an advanced state of completion. Unfortunately, owing to the war, building operations ceased, and were not resumed again until 1920.

Aims and Objects of the National Museum.

His Majesty King George V., in laying the foundation stone, defined the objects of the Institution in the following terms :—" The collections in this Museum will serve as a record of the development of every branch of intellectual and industrial activity, and will illustrate the practical aspects of Welsh life." The aims which have inspired the work of the Council have coincided with His Majesty's definition, and were concisely expressed in the loyal address, presented by the President to His Majesty on the above-mentioned occasion, as " to teach the World about Wales, and to teach

SEUM OF WALES.

The National Museum of Wales.

the Welsh people about their own Fatherland " (" i ddysgu'r byd am Gymru, a dysgu Cymry am Wlad eu Tadau ").

In the scheme which has been prepared, the Council hopes to give visitors to the Museum a full and true presentment of Wales from the natural, the historic, and the artistic standpoints. It has organised departments representing, on the one hand Geology, Mineralogy, Zoology, and Botany, and on the other, national Ethnography, Archaeology, History, and Art. To these are being added a library of books of reference dealing with subjects that come within the scope of the Museum, for the use of students and specialists, a lecture theatre for the purpose of explaining the contents of the Museum to popular audiences, and a department to arrange loan collections for local Museums, Exhibitions or Eisteddfodau throughout Wales. The Museum will rank as one of the most important educational influences of the Principality, for by its collections specially arranged to meet the needs of Investigators, Students, and School Children, as well as by its loan collections and its library, it will be the natural complement and help-meet of the University of Wales, with its constituent Colleges, the School of Mines, Treforest, numerous Scientific and Artistic Associations and Schools, the National Memorial to King Edward VII., and other similar institutions.

The Museum will be no mere collection of curiosities, but an institution designed to fulfil the functions performed in Scotland by the four State-aided institutions, the Royal Scottish Museum, the National Gallery of Scotland, the Scottish National Portrait Gallery, and the Museum of the Society of Antiquaries of Scotland ; and in Ireland by two State-aided institutions, the National Museum of Science and the National Gallery of Ireland ; and consequently the building to be provided for it must be on a considerable scale, and capable of enlargement on a definite plan. The external appearance of the building may be seen from the accompanying illustration. The internal arrangements have been dictated by definite administrative principles. The first of these is the clear division of the contents of the

Museum into two main classes: The show collections intended for the ordinary visitor, and the reserve or study collections intended for the specialist and the student. The exhibition galleries will be devoted entirely to the former, and not utilised as store-houses, whilst the latter will be kept in special rooms adjoining the exhibition galleries. The effect of this will be that those who wish to investigate quietly, will do so in a suitable room, with tables, microscope, and other apparatus at their disposal, undisturbed by the passing throng, whilst the visitors will not find their view of the cases blocked by students. The reserve galleries will be the proper place for shells and beetles systematically arranged. This system will admit of those specimens which are selected for exhibition to the public being displayed in such a way as to exhibit their characteristics to the best advantage, and, in the case of works of art, of their having that amount of space which is necessary for them to produce their proper effect.

It is intended that the New Museum shall be primarily and essentially National in character. It is impossible, however, thoroughly to understand either the natural or artificial products or the history of Wales without examples drawn from a wide area for comparison. Nothing should be admitted, however, which does not illustrate or elucidate Wales in some aspect or other. Speaking generally, it is intended that there shall be a small general collection connected with each department of the Museum; for example, at the beginning of the Zoological gallery there will be a small type series giving a general survey of the Animal Kingdom, which will explain the special relationships of the animals of Wales. It is also intended to form a complete collection of Fossils and other Geological specimens to illustrate points of interest to Miners and students of questions connected with the Coal Industry. Such a collection would be a most valuable aid to the University Colleges of Wales and to the various Technical Institutions in the Principality.

The natural history of Wales (zoology and botany) is to receive special illustration in a hall designed for this particular purpose, in which groups of animal and plant life are to be installed in assemblages such as they form in Nature. For this purpose it is proposed to arrange a series of large cases or small rooms, each containing, as it were, a sample of some part of the country—moorland, forest, woodland, pond, marsh, and shore. Each will have a suitable background, which may include an actual Welsh landscape, and in the foreground will be real or artificial plants and mounted animals as realistic as the resources of modern taxidermy can make them. A descriptive label, coupled with a diagram, will give the names of the animals and plants, and show their relationship to their environment.

A similar method is to be adopted to illustrate and explain Welsh history. To begin at the very beginning, it is intended to have a mounted group showing the prehistoric man sitting at the entrance of his cave chipping stone implements. Another, illustrating the Roman period, might contain a reconstruction of the gateway of a Roman camp, with the sentry standing on duty, and natives of the country buying and selling produce. Ample materials exist to furnish the basis of such a restoration. In a similar way, exterior or interior groups will show subsequent periods; a Welsh castle of Norman times, a Tudor mansion, and a Welsh farm of the last century will all be illustrated with the appropriate furniture, fittings, and utensils.

As regards applied art, it seems feasible to include some of the best examples of the work of other lands and ages. A series of cases may be arranged, one devoted to Egyptian, another to Oriental, and a third to mediæval art, whilst examples of first-class painting, engraving, and sculpture will be placed in other galleries to train the æsthetic instincts of students and the public.

The Present Position.

But for the war, the building as it is seen to-day would have been completed and occupied some years ago. As it is, the contractors are still in possession, and owing to the greatly increased cost of building at the present time, the Council only sees its way to proceed with the completion of the interior of the western section. To do merely this, considerable financial help is still required. Much work, however, is being accomplished by the staff of the various Departments in making ready for exhibition and study the large and important collections which the Museum possesses. Temporary exhibitions are organised from time to time, lectures given up and down Wales by the members of the staff, and loan collections are distributed. The public generally makes much use of the expert knowledge available in the different sections, and parties of students pay frequent visits to see collections which it is not possible to show permanently for want of accommodation.

The Authorities of the Museum are sparing no effort to expedite the opening of the new building and to ensure that the Institution shall take its proper place in the educational system of Wales.

The present officers of the Museum are The Right Hon. Lord Treowen, C.B., C.M.G. (President), Major David Davies, LL.D., M.P. (Vice-President), Alderman Illtyd Thomas, J.P. (Treasurer), with Dr. William Evans Hoyle (Director).

" The Cardiff Collections " in the Museum, Trinity Street, are open to the public at specified times. They will be open from 10.0 to 12.30 and from 2.0 to 5.0 each day during the meeting of the British Association.

A temporary exhibition of specimens will be open during the same hours in the Museum Room, City Hall.

The New Building may be inspected by members of the British Association on application to the Clerk of Works or the Foreman on the site.

RAILWAYS.

By THOS. A. WALKER.

It has been said with a great deal of truth that the "History of the Railways is the History of the Coal Trade";* but in tracing the origin of the railways in the area of the South Wales Coalfield, it is necessary to go somewhat further back, and consider the establishment of the great Iron Industries that were set up in the different valleys of this hitherto quiet and rural district.

The presence of iron in these valleys was known from very early times, and it was produced by the farmers by smelting with charcoal in little "bloomeries" (holes dug in the soil) while the blast was made by means of portable bellows. The refuse heaps, which remain, indicate that these works were very small, scarcely more than a ton of iron being smelted at a time.

The first serious attempt to produce iron in Wales was made at Dowlais in 1758, when William Lewis set up a small furnace and engaged as his ironmaster John Guest, a small freeholder of Broseley, in Staffordshire. Iron making in Staffordshire had been in repute for many years, where again the smelting of the ore was accomplished by the use of charcoal, the water-wheel having been introduced to keep up the blast. This method was followed at Dowlais for a considerable period, until the clearing of the hillsides of timber made the supply of charcoal difficult to maintain. It was then resolved to use coal instead. That this was a satisfactory solution subsequent events have amply proved. From this time on the success of the iron trade at Dowlais was assured, and the present works are amongst the largest in the country.

Other outstanding figures in the early history of the iron trade are Anthony Bacon, who founded the Cyfarthfa Works in 1763, and the Homfrays, who became his managers in

* E. A. Pratt, "History of Inland Transport,"

1782. This connection lasted for two years only and then ensued a quarrel and a separation, which resulted in the establishment by the Homfrays of another ironworks at Penydarren, a dingle near Morlais, which they rented for £3 per annum. Having succeeded in obtaining means to finance their undertaking, and in building up a large and prosperous business, they extended their field of operations beyond the Merthyr district, and set up the first furnaces at Ebbw Vale, Sirhowy, and Tredegar.

The greatest difficulty was encountered in getting the output away, as the roads in the district were infamous. "They were mere tracks over the mountains, or cartways deep in mud that formed quagmires, and if metalled, were so, with boulders and unbroken stones."* The difficulty was overcome, chiefly owing to the efforts of Anthony Bacon, by the making of a road to Cardiff. This was completed in 1767, the cost of it being defrayed by public subscription. By this means the iron was transported to the sea, sometimes by carts, and very frequently on the backs of mules.

This primitive method of conveyance continued for some thirty years, when the necessity of improved methods of transport in the interests of the ironworks, and of the collieries which had come into existence in the district, led to the promotion of a scheme for a canal between Merthyr and Cardiff, an Act for which was obtained in 1790, by the Company of Proprietors of the Glamorganshire Canal Navigation. Amongst the names of the promoters are those of William Crawshay (who succeeded Anthony Bacon at Cyfarthfa), Jeremiah and Samuel Homfray, and Thomas Guest. This canal was opened in February, 1794, and is described by I. Phillips as having "opened a ready conveyance to the vast manufacture of iron established in the mountains of that country, and many thousands of tons are now annually shipped from thence."†

Other Canals in the South Wales area were the Monmouthshire canal, from Newport to Pontypool with a branch to

---

* Baring Gould, "Book of South Wales.'

† "General History of Inland Navigation, 1803."

Crumlin, the Aberdare Canal from Aberdare to the basin of the Glamorgan Canal at Navigation (now called Abercynon), the Neath Canal from Pont Walby (Glyn Neath) to Neath, the Tennant Canal from the Neath River to Swansea, and the Swansea Canal from that port to Abercrave.

The opening of the canals resulted in a development of railways in South Wales, as in other places, to supplement the canals. These railways could be used by anyone, in connection with canal transport, on payment of the stipulated tolls. As a matter of fact it was in this district that the early railways eventually attained their greatest development. Writing in 1824, T. G. Cumming, Surveyor, Denbigh, states :—" As late as the year 1790 there was scarcely a single railway in South Wales, whilst in the year 1812 the railways, in a finished state, connected with canals, collieries, iron and copper works, etc., in the Counties of Monmouth, Glamorgan, and Carmarthen alone, extended to upwards of one hundred miles in length."* By the year 1830 this was increased to about 350 miles. The whole of these railways were on the tram-plate or flanged rail principle, solid blocks of stone being used in Wales, instead of the wooden sleepers common to most other districts.

The advisability of constructing railways, or tram roads, in order to connect the canals with particular works in their neighbourhood, had assumed such importance in the minds of the undertakers of the canal navigations, that it became customary towards the end of the eighteenth century for the canal companies in applying to Parliament for powers, or extensions of existing powers, to seek authority themselves to make railways, or tramways, in connection with their undertakings. In the event of the canal company not exercising these powers, the adjoining landowners whose estates contained any mines of coal, iron-stone, lime-stone, or other minerals, or the proprietors of any furnaces or other works lying within four miles (in some cases eight miles) of some part of the canal, were empowered to carry

* " Illustrations of the Origin and Progress of Rail and Tram Roads and Steam Carriages or Locomotive Engines "

out the work themselves at their own cost without the consent of the owner of the lands, rivers, brooks, or water-courses it might be necessary to cross, on payment of such similar compensation as would be due in regard to the construction of canals. Section 57 of the Act for the making of the Glamorgan Canal is a typical example.

The iron trade had, however, developed so rapidly, that although the canal from Merthyr to Cardiff had been completed, and boats arrived daily at the latter place laden with the produce of the iron works, it failed to meet the iron merchants' requirements. Further, the upper part of the canal appears to have suffered from the want of water in summer, and to have been frequently frozen over in the winter. A scheme, therefore, for a railroad between the points was projected in the first year that the waterway was opened; an Act of Parliament being obtained by Samuel Homfray, and others, for constructing an iron railway between Merthyr and Cardiff that should be free for any persons to use with trams of the specified construction, on paying certain tonnage, or rates per mile, to the proprietors. This project was abandoned for some time, but was revived in 1799, when a more ambitious scheme was put forward. The *House of Commons Journals* record that on February 18th of that year, William Lewis (Alderley), William Tait, Thomas Guest, Joseph Cowles, and John Guest, being a firm of ironmasters in the Parish of Merthyr Tydfil, known as the Dowlais Iron Company; Jeremiah Homfray, Samuel Homfray, Thomas Homfray, and William Forman, ironmasters of Merthyr Tydfil, known by the name of Jeremiah Homfray & Co., Richard Hill (Plymouth Works), and William Lewis (Pentyrch Works), petitioned the House for leave to bring in a bill for the construction of a "dram road;" from or near Carno Mill, in the parish of Bedwellty and the County of Monmouth, to Cardiff, with branches to Merthyr and Aberdare. Carno Mill is situated at Rhymney, where another iron works had been established.

For some reason, probably the opposition of the Canal Company, this Bill did not reach a second reading, but a line was made that constitutes a landmark in early railway history and from the peculiar circumstances under which it was constructed, is undoubtedly unique in the annals of railway enterprise. The line in question extended from Dowlais and Merthyr to the Canal Wharf at Navigation (Abercynon), and was actually made under the powers of Section 57 of the Glamorganshire Canal Company's Act. This clause limits the length of any tramway constructed under its provisions to four miles, but the promoters of the tramroad put forward as their contention that so long as their mines or works were within four miles of the canal, they had the right to lay down their tramroad to such a point on it as they might select, and chose Navigation as the place best suited to their purpose.

When the project was completed, a railway was made, which was of a total length of nine miles from Merthyr and ten from Dowlais. It was on this line that the first steam locomotive ran on February 14th, 1804. An engine built by Richard Trevithick hauled a load of iron from J. Homfray's works to Navigation at the rate of five miles per hour. The experiment was, however, not a success, as in cousequence of the gradients and curves of the line the engine failed to bring the empty trams back to the works.

The iron and coal trade of the district was by this time making great progress, and railways were made both by the iron masters and the colliery owners, in order to dispose of their produce ; and also by the adjacent land owners for the sake of the tolls which would accrue to them for the use of their lines. The result was that networks of railways sprang into existence, linking up the works and collieries with the canals, and in one case connecting the waterways themselves. Some of these railways were considerable undertakings, as, for instance, the Sirhowy Tram Road, which was made in part under the powers of the Monmouthshire Canal Act, but for its greater length under an Act passed in the 42nd year of the reign of George III., the Royal Assent being obtained on June 26th, 1802.

This line of tramway, apart from the lines constructed to connect with canals, can be considered as the earliest railway in South Wales. This Act was obtained conjointly by the Monmouthshire Canal Company, Sir Charles Morgan, and the Tredegar Iron Works Company. The Canal Company's Act (32 George III., 1792) gave authority to the canal undertakers to make railways within a distance of eight miles from their canal. This power of the Canal Company was combined with the powers granted to the Tredegar Company, and with those obtained by Sir Charles Morgan to make the line through his Park, resulting in a line of railway from the Port of Newport to Tredegar, 28 miles in all, being completed, and opened for traffic in August, 1811.

The Rumney Tramway appears to have been promoted under somewhat different circumstances, as the whole of the money required to make this line, about £48,000, was subscribed by four persons, Sir Charles Morgan, Joseph Bailey, William Thompson, and Crawshay Bailey. The Act, 6 George IV., Cap. 62, received the Royal Assent 20th May, 1825, and is described as an Act for making and maintaining a railway or tramway from the northern part of a certain estate called Abertysswg, in the parish of Bedwellty, County of Monmouth, to join the Sirhowy Railway at or near Pye Corner in the parish of Bassaleg in the same County. Besides serving the Rhymney Iron Works this line was also used for conveying limestone from the quarries at Machen. It was 21 miles and 6 furlongs in length, and the total rise from end to end 756 feet. Locomotive engines were employed on this line between 1840 and 1845, and it continued to be worked as a mineral line until 1863, when it was taken over by the Brecon and Merthyr Railway and converted into a public passenger railway.

Other railways or tramways were made under the powers of the Monmouthshire Canal Act, which to-day constitute the present Eastern and Western Valleys system of the Monmouthshire Section of the Great Western Railway.

The total mileage of tramways or railroads of 4 feet

4 inches gauge in South Wales, constructed under the authority of Parliamentary powers of the Canal Acts or tramroads apart from canals, within the period of 1792 and 1830, amounts to some 347 miles of single line, and is given in detail in the following table.

Name	Act of Parliament	Miles.
Glamorgan Canal	30 Geo. III. June 9th, 1790 ..	25
Monmouthshire Canal	32 ,, ,, 3rd, 1792 ..	40
Carmarthen Canal	42 ,, ,, 3rd, 1802 ..	16
Sirhowy Tramway	42 ,, ,, 26th, 1802 ..	28
Oystermouth Railway	44 ,, ,, 29th, 1804 ..	12
Neath Canal and Railway	31 & 38 ,, 6th, 1798 ..	6
Kidwelly Canal and Railway	52 ,, ,, 20th, 1812 ..	10
Monmouth Railway	50 ,, May 24th, 1810 ..	25
Llanfihangel Railway	51 ,, ,, 25th, 1811 ..	10
Penclawdd Canal and Railway ..	51 ,, ,, 21st, 1811 ..	2
Grosmont Railway	52 ,, ,, 20th, 1812 ..	8
Hay Railway	51 ,, ,, 25th, 1812 ..	25
Kington Railway	58 ,, ,, 23rd, 1818 ..	14
Mamhilad Railway	54 ,, June 17th, 1814 ..	6
Hereford Railway	7 Geo.,IV., May 26th, 1826 ..	14
Rumney Railway and Tramway ..	6 ,, ,, 20th, 1825 ..	22
Duffryn, Llynvi and Dock	6 ,, ,, 10th, 1825 ..	17
Dulais	7 ,, ,, 26th, 1826 ..	7
Bridgend	9 ,, June 19th, 1828 ..	4½
Llanelly and Dock	9 ,, ,, 19th, 1828 ..	2½
Llanarth Tramway	.. ..	3
Hall's Tramroad	.. ..	8
Cwm Dows and Kendon Tramroad ..	.. ..	3
Tal-y-bont and Bryn Ore Co.	.. ..	11½
Rhymney Iron Company's Tramroad	.. ..	3½
Black Forest, Brecknock, to Sennybridge	.. ..	10
Penwyllt to Ystradgynlais, &c ..	.. ..	14
		347

E. A. Pratt points out " that whilst the iron masters and owners of mines, coal, etc., introduced these tramways in the first instance, it was the canal companies themselves who in the days before locomotives, mainly developed and established the utility of a new mode of traction which was eventually to supersede to so material an extent the inland navigation they favoured."*

This method of transportation continued until the year 1830, with fairly good results, but from this time on it was evident that it was not competent to cope with the ever-increasing output of the various works, and the owners of the Dowlais Works, and others, were desirous of finding a

* " History of Inland Transport "

quicker method of transportaion for their manufactures to Cardiff. The result was a meeting of the proprietors of iron works, collieries, and others interested in the mineral and other property of the Valleys of the Taff, Rhondda, Cynon, Bargoed, and other adjacent places, together with those interested in the trade of the town of Merthyr Tydfil and Port of Cardiff. This meeting, presided over by John Guest, M.P., was held on the 12th October, 1835, at the Castle Inn, Merthyr, in order to take into consideration the improvement of the communication between these places and the Port of Cardiff. Resolutions were passed to the effect that the present means of communication did not afford the requisite facilities for transporting to the sea the productions of the various works and collieries in these districts, and that it was therefore "expedient to establish a communication by means of a railway which should combine the advantages of the latest improvements in the mode of transport."

For this purpose the TAFF VALE RAILWAY COMPANY was formed, and in 1836 a Bill was introduced into Parliament for the making of a railway from Merthyr Tydfil to Cardiff, to be called the "Taff Vale Railway." The principal opposition to the Bill came, as might have been expected, from the Glamorganshire Canal Company which, up to this time, had a monopoly of the coal and iron traffic sent from the various collieries and iron works to Cardiff, but an arrangement including the payment of a lump sum seems to have been made by which all further opposition to the Bill was withdrawn. During the Parliamentary proceedings another arrangement was made with the proprietors of the Dowlais, Penydarren, and Plymouth Iron Works, for the purchase of the tramroad from Merthyr to Abercynon in order to release the three Iron Companies from a deed of covenant under which they were bound to pay tonnage upon the old tramroad whether they used it or not.

The Act for making the line was obtained on the 21st June, 1836, and the first section from Cardiff to Navigation (Abercynon) was opened for traffic on October 17th, 1840, and to Merthyr Tydfil on April 5th in the following year. This was the first railway constructed on modern lines in South Wales and was a single line, 24½ miles in length, provided with six passing places. There were two short branches, one of which ran up the Rhondda Valley to a point near Dinas, where a junction was effected with the tramroad giving access to Walter Coffin's colliery at that place; and another to Llancaiach in order to reach the collieries situated at Tophill and Gelligaer. A further branch was authorised in the Act which was to give access to a "good port" the site of which was fixed at the mouth of the River Ely near Penarth. By an agreement entered into with the Marquis of Bute the proposed branch railway was, however, abandoned in return for a lease of the east side of the West Bute Dock, Cardiff, for 250 years.

Contrary to expectation, the railway for some years was not pecuniarily successful, but it subsequently made great progress and developed from a poor paying concern to one of the best dividend earning railways in Great Britain.

The opening of the Aberdare Railway, which was incorporated by an Act of Parliament of July 31st, 1845, and was leased in perpetuity to the Taff Vale Company from 1st January, 1847, gave renewed vitality to the undertaking, the success of which was further assured by the extension of the railway in the Rhondda Valley to Treherbert. The powers for this extension were obtained in 1846, but the line was not opened throughout for passenger traffic until January 7th, 1863. Coal traffic had, however, been worked from Lord Bute's Colliery at Treherbert as early as the year 1855. The following table will give an idea of the development of the coal traffic during this period:—

Coal and Coke Traffic conveyed over Taff Vale Railway.

	Tons.
Year 1841	41,669
„ 1850	594,222
„ 1860	2,132,995

The year 1845 saw tremendous activity in the promotion of railways, no fewer than 815 companies in all parts of the country were formed, and their plans laid before Parliament. South Wales was no exception to the rule, and during that and the following year (1846) five lines were authorised which, when completed, had a considerable effect on the Taff Vale Railway by affording railway access to all parts of the United Kingdom. These Railways were :—

1. The Oxford, Worcester and Wolverhampton Railway (Incorporated 1845).
2. South Wales Railway (Chepstow to Swansea) (Incorporated 1845).
3. Newport, Abergavenny and Hereford Railway (Incorporated 1846).

   Ditto (Taff Vale Extension) (Incorporated 1847).
4. Vale of Neath Railway (Incorporated 1846).
5. Llynvi and Ogmore Railway (Incorporated 1846).

(All these lines are now integral parts of the Great Western Railway.)

Under the powers of the Newport, Abergavenny, and Hereford Railway (Taff Vale Extension) Act, 1847, that Company was authorised to make a junction with the Taff Vale Railway Company at Quakers Yard. This was completed and brought into use in January, 1858, and provided the first opportunity the Taff Vale Company had of exchanging vehicles with another railway company. This was of great importance, as it afforded access to the Midlands and South Staffordshire. It is true that another means of exchanging traffic existed at Cardiff, where a connecting line was made by the South Wales Railway, but owing to the difference in the gauges of the two railways, it was impossible to exchange the vehicles, with the result that a third rail had to be laid which enabled the Taff Vale Company to work over the junction to what was at that time the Goods Station of the South Wales Railway. Here goods (including coal) were transhipped from one vehicle to another.

The continued increase in the output of coal in the Aberdare and Rhondda Valleys led to a revival of the scheme for making a railway from the Taff Vale Railway to the River Ely, and for converting part of the river into a tidal harbour.

For this purpose a Bill was deposited in Parliament by a separate Company, entitled the Ely Tidal Harbour and Railway Act, 1856, on the ground that it "would be of great public and local benefit." The following gentlemen were incorporated as subscribers:—The Hon. Windsor Clive, Crawshay Bailey, Thomas Powell, Rev. George Thomas, William S. Cartwright, John Nixon, James Harvey Insole, and others.

This Bill was opposed by the Taff Vale Company, but was subsequently approved by the Shareholders at a Special Meeting convened for that purpose, consequent upon the insertion of protective clauses and a special clause as to traffic arrangements. In the following year further powers were obtained to construct a dock and the necessary railways and works on the south-west bank of the River Ely, the name of the Company being altered to "The Penarth Harbour, Dock and Railway." In 1863, a further Act was obtained authorising the Penarth Company to lease their undertaking to the Taft Vale Company for a term of 999 years.

That the undertaking was of "great public and local benefit" is confirmed by the shipments of coal at the Dock since the opening in 1865, viz.:—In 1866, the first complete year after the opening, the tonnage was 572,404, and in 1907, the heaviest on record, it had risen to 4,574,974 tons.

The Taff Vale Railway to-day has a total mileage of 124 miles and 42 chains, and in addition to the lines constructed under the Acts promoted by the Company, has acquired the following undertakings:—

	Incorporated.	Opened.	Leased.	Amalgamated.
Aberdare Railway	1845	1846	1848	1902
Penarth Harbour, Dock and Railway	1856 1857	1865	1863	..
Llantrisant and Taff Vale Junction Railway	1861	1863	..	1889
Cowbridge Railway	1862	1865	..	1889
Dare Valley Railway	1863	1866	1870	1889
Rhondda Valley and Hirwain Junction Railway	1867	1878	..	1889
Penarth Extension Railway ..	1876	1878	1877	..
Treferig Railway	1879	1883	1884	1889
Cardiff, Penarth, and Barry ..	1885	1887	..	1889
Cowbridge and Aberthaw	1889	1892	..	1894

(In addition, two privately owned lines were purchased by the Taff Vale Company. Ferndale to Maerdy in 1886, and Clydach Vale Colliery Railway in 1899.)

The GREAT WESTERN COMPANY'S system of railways in the area of the South Wales Coalfield under notice in this article is composed of nineteen different undertakings, which have been either leased to, or amalgamated with, the Company on various dates, and this combination now forms one of the busiest and best paying portions of the Great Western Railway, both in respect of mineral and passenger traffic. Taken in their chronological order they are as follows :—

	Incorporated		
Duffryn, Llynvi, and Porthcawl Railway	1824	Consolidated with Llynvi Valley Railway	1847
South Wales Railway Chepstow to Swansea	1845	Amalgamated with G.W. ..	1863
Llynvi Valley Railway	1846	Amalgamated with Llynvi and Ogmore	1866
Vale of Neath Railway	1846	Amalgamated with G.W. ..	1866
Newport, Abergavenny, and Hereford Railway	1846	Amalgamated with West Mid.	1860
Ditto, Taff Vale Extension ..	1847		
Briton Ferry Railway and Dock	1851	Amalgamated with G.W. ..	1873
Coleford, Usk, Monmouth, and Pontypool Railway ..	1853	,, , .	1881
Llanelly Railway and Dock ..	1853 (as a reconstructed Company)	,, ,, ..	1889
Aberdare Valley Railway ..	1855	Leased or sold to Vale of Neath	1859
Bristol and South Wales Union Railway	1857	Amalgamated with G.W. ..	1868
Ely Valley Railway	1857	Leased to G.W.	1862
Ditto Mwyndy Branch ..	1858		
Swansea and Neath Railway ..	1861	Amalgamated with Vale of Neath	1863
Ogmore Valley	1863	Amalgamated with Llynvi and Ogmore	1866
Ely Valley Extension Railway..	1863	Amalgamated with Ogmore Valley	1865
Pontypool, Caerleon, and Newport	1865	Amalgamated with G W . ..	1876
Llynvi and Ogmore Railway ..	1866	.	1873
Severn Tunnel Railway ..	1872	.. .	1872
Ely and Clydach Valleys Railway	1873	,, ,, ..	1880
East Usk Railway (Newport) ..	1885	,, ,, ..	1892

The West Midland Railway was amalgamated with the Great Western in 1863.

The Monmouthshire Section of the Great Western Railway, which forms so considerable a portion of that Company's system in South Wales has a long and very interesting history. Authority was obtained from Parliament as far back as 3rd June, 1792, for making and maintaining a navigable canal from a place near Pontnewynydd into the River Usk, at or near the town of Newport, and for constructing a collateral canal from the same, at or near a place called Crindau Farm to Crumlin Bridge, an ancient structure carrying the public road over the river at that place. Powers were also obtained for making a railroad from the ironworks at Blaenavon to the canal at Pontnewynydd, and a similar railroad from the canal at Crumlin to the ironworks at Ebbw Vale, Beaufort, and Sirhowy, with a branch to Nantyglo and Brynmawr. These powers were extended by a second Act passed in 1797, and again by a third Act, which was obtained in 1802. As in the case of the Glamorganshire Canal, this method of transportation proved insufficient for the growing trade of the district, and in 1845 an Act was obtained to make a railway from Newport to Pontypool, and to enlarge the powers relating to the Company. The provisions of this Act were amended in 1848, and the name of the Company was altered from "The Company of Proprietors of the Monmouthshire Canal Navigation" to the "Monmouthshire Railway and Canal Company." Under these subsequent Acts, railways were made from Trosnant to Pontypool; from Cwmbran, Llantarnam and Ponthir, to the River Usk at Caerleon; from the junction with the Sirhowy Railway at Risca to Crumlin, with branches in various directions to reach the numerous works in the vicinity. In the year 1880, in order to contribute to the economical and efficient working of the Monmouthshire undertaking and to the public convenience, powers were sought to amalgamate the Company with the Great Western Company. The Act by which this amalgamation was brought about obtained the Royal Assent on the 2nd August, 1880.

At the present time the various railways in South Wales amalgamated or leased to the Great Western Railway, including the South Wales line from Chepstow, have an approximate mileage of 550 miles.

The BRECON AND MERTHYR TYDFIL JUNCTION RAILWAY COMPANY was incorporated by Act of 1st August, 1859, for constructing railways in the district between Brecon and Merthyr Tydfil. By this Act the Company were authorised to make railways from Merthyr to Talybont and to effect a junction with the Dowlais Railway constructed under the powers of "The Dowlais Railway Act, 1849." The Act also gave authority for the construction of a "Stone Road" to Llansantfraed, to be carried across the River Usk by a bridge. A further Act was obtained in 1860 by which the construction of this last undertaking was relinquished and powers obtained for the extension of the railway to Brecon and for the making of a junction with the Mid-Wales Railway at Talyllyn, thus opening up a route to Liverpool, Birkenhead, and Manchester, by way of Oswestry, Newtown, Llanidloes, and the Valley of the Wye, from Swansea, Cardiff, and Merthyr Tydfil. The Act authorised the raising of £50,000 new capital, to which the Taff Vale Company might contribute £30,000. In 1861, the Brecon and Merthyr Company were authorised to extend their line from Pant to Aber Bargoed, at which place they were empowered to make a junction with the main line of the Rhymney Railway Company.

This line was only completed as far as Deri, the Rhymney Company having obtained powers to make a branch from their main line to or near that place, and here the junction between the two Companies was made. In the following year authority was obtained to make a junction with the Taff Vale Railway at Merthyr. By the Rumney and Brecon and Merthyr Railways Act, 1863, authority was given for the sale of the Rumney Tramway to the Brecon and Merthyr Company. By this transaction the latter Company obtained access to Newport. The Rumney Company had been re-incorporated in August, 1861, with authority to alter the line and level of the existing tramroad, and to make a new railway from Machen Upper to the Caerphilly Branch of the Rhymney Railway.

The RHYMNEY RAILWAY COMPANY came into existence in 1854. In July of that year an Act was obtained to make a railway from Rhymney to Hengoed, where a junction was made with the Great Western Railway (Newport, Abergavenny, and Hereford section) with a branch from Bargoed into the Rhymney Valley. The line was promoted by John Boyle, Henry Austin Bruce, John Daniel Thomas, Jonathan Worthington, and John Nixon, and power was taken to enable the Trustees of the Marquis of Bute (O. T. Bruce and John Boyle) to subscribe Trust money in the undertaking, on the ground that the formation of the railway would be highly beneficial to the estate of the Marquis of Bute. In 1855 the Company was authorised to extend the main line from Hengoed Junction to a junction with the Taff Vale Railway at Walnut Tree Bridge, with running powers over the latter Company to Cardiff; to make a branch line to Caerphilly; and branches at Cardiff to the Bute East Dock and Tidal Harbour, leading from and out of the Taff Vale Railway. By this means a through communication was afforded between Rhymney and Cardiff and Penarth Docks. This arrangement continued until 1864, when an Act was obtained by the Rhymney Company authorising the construction of a direct line from Caerphilly to Cardiff, thus obviating the use of running powers over any other company's system. Accordingly the powers to run over that portion of the Taff Vale Railway between the Penarth and Crockherbtown Junctions ceased on the opening of the direct line on 1st April, 1871. By subsequent Acts the Company was enabled to effect important extensions and junctions with the systems of other Companies, thereby establishing through routes from Cardiff and Rhymney to all parts of the West and North of England, and to Merthyr, Aberdare, and the Vale of Neath in South Wales.

The LONDON AND NORTH WESTERN RAILWAY COMPANY have considerable sections of railway in this area. Of these, the Merthyr, Tredegar, and Abergavenny Section, with its subsidiary branches, is the most important. This railway was incorporated as a separate undertaking by Act 1st August, 1859, to supply railway communication to the district between Merthyr and Abergavenny. By an Act passed in August, 1862, the line was leased for a thousand years to the London and North Western Company, and in 1866 the Company was dissolved, the line now forming part of the London and North Western System.

The Sirhowy Railway was, as before stated, incorporated in 1792 as the "Sirhowy Tramroad Company." It was altered and modernised in 1860, when the name was changed to the Sirhowy Railway, further powers being obtained in 1865. The line extends from the Merthyr, Tredegar, and Abergavenny section of the London and North Western system, near Nantybwch Station, to Tredegar, and thence to Nine Mile Point, where it joins the Monmouthshire section of the Great Western Railway. It traverses an important coalfield, and continued as a separate undertaking until the 13th July, 1876, when it was vested in the London and North Western Railway.

The ALEXANDRA (NEWPORT AND SOUTH WALES) DOCKS AND RAILWAY was incorporated in 1865 as the Alexandra (Newport) Dock Company. The Company was authorised to construct docks, railways, and other works in connection therewith, all of which were opened in April, 1875. The name of the Company was altered to the above title in 1882. By an Act of 1897 the Alexandra Company acquired the Pontypridd, Caerphilly, and Newport Railway, which latter Company was incorporated in 1878 for the construction and maintenance of a railway commencing with a junction with the Taff Vale Railway, near Pontypridd Station, and terminating by a junction with the Caerphilly branch of the Rhymney Railway at Penrhos, from which point the Pontypridd Company had running powers to Bassaleg Junction over the Rhymney and Brecon and Merthyr lines respectively. By subsequent Acts in 1883 and 1887 powers were obtained to construct a new line from Bassaleg Junction to Alexandra Docks. By means of this line the coal traffic from the Rhondda, Aberdare, and Merthyr Valleys is conveyed direct to Newport for shipment. In 1911, powers were granted to the Alexandra Company to make a railway from the Tredegar Park Mile Railway to a junction with the Sirhowy Branch of the London and North Western Railway at Nine Mile Point, but the line was not constructed.

The RHONDDA AND SWANSEA BAY RAILWAY was authorised by Act of 10th August, 1882, to connect the Rhondda

Valley coalfields with the Swansea Bay ports. The line commences with a junction with the Taff Vale Railway at Treherbert, running through the new coalfield in the Avon Valley, and terminating at Briton Ferry, with a branch line to Cwmavon, Aberavon, and Port Talbot Docks. In August, 1883, the Company obtained further Parliamentary powers to extend their line from Briton Ferry to the Prince of Wales Dock at Swansea. The line was opened from Aberavon into the Rhondda Valley on the 2nd July, 1890. The section from Aberavon to Briton Ferry was opened for mineral traffic 30th December, 1893, and the line thence to Swansea on 14th December, 1894.

The BARRY RAILWAY COMPANY was incorporated by Act of 14th August, 1884, for the construction of a dock at Barry Island, seven miles from Cardiff, and of railways from the Dock to the Rhondda Valley, with access by junctions with existing and authorised railways, to all the other great mineral producing districts of the South Wales Coalfield. Further powers were obtained in 1896 to construct railways from or near St. Fagans to points on the Walnut Tree and Aber Branches of the Rhymney Railway, and in 1898 authority was given for the construction of a line from the Company's Aber Branch to join the Brecon and Merthyr Railway near Bedwas, with a short branch to effect another junction with the Rhymney. In August, 1907, powers were obtained for railways from the Company's Penrhos Junction to the Sirhowy Branch of the London and North Western and the Western Valley Branch of the Great Western, but these lines have not yet been made.

The Company, under agreement, work the Vale of Glamorgan Railway, which was authorised on 26th August, 1889. This line commences with a junction with the Barry Railway at Barry, and terminates at a junction with the Llynvi and Ogmore Section of the Great Western at Coity, and has three short branches.

The Barry Company exercise running powers over the Taff Vale and Great Western Railways to Cardiff for passenger traffic, and also over the latter Company from Cardiff to St. Fagans for similar traffic to Pontypridd.

The last Railway Company to be mentioned in this article is the Cardiff Railway Company. On August 6th, 1897, the Bute Docks Company were granted powers to construct certain railways in order to obtain improved railway access from the Docks of the Company to the Taff and Rhondda Valleys, and other portions of the Glamorganshire Coalfield, and at the same time obtained permission to alter the name of the existing Company to that of the Cardiff Railway Company. This scheme provided for the making of five short lengths of railway, aggregating about 12 miles in length, and the making of junctions with the Rhymney and Taff Vale Railways at or near Cardiff, and also with the Taff Vale Railway at Treforest. The line has been completed in part and is open for passenger and mineral traffic as far as Rhydfelin, but the junction with the Taff Vale Railway has not yet been put in.

Briefly summarised, the position is that during the last century, nearly 1,600 miles of tramroads and railways have been constructed in South Wales. Of these, 350 miles were laid down in the interval between 1795 and 1840. This year may be taken as the close of the tramroad era, and since that time 1,250 miles or thereabouts of modern passenger and mineral lines have been either constructed, or converted, with the result that this area of Wales is now one of the busiest and most densely populated of the United Kingdom.

DOCKS.

By W. J. HOLLOWAY.

ALTHOUGH Cardiff may claim to be one of the earliest of the Welsh ports, and possibly the pioneer Docks of the coal trade, she cannot claim, as Bristol does, any great antiquity. Before Cardiff was thought of as a Port, vast fortunes were acquired by Bristol merchants, who, in the palmy days of the slave trade, practically controlled the rich traffic of the West Indies. At the time referred to, the Welsh side of the Bristol Channel was noted more for its pirates than for its trade, and if all the stories handed down of the piracy practised by the notorious "Will of the Iron Hand" in the neighbourhood of Dunraven are true, possibly some of the ships passing to and from Bristol, with their rich cargoes, were lured to their doom on that wild rock-bound coast. Happily that state of affairs was not allowed to continue, and later, when rich veins of coal were discovered in the valleys of Wales—coal of unrivalled quality and almost unlimited in quantity—men were forthcoming, who were determined to build up what has now proved to be a great national asset in the expansion of the industries and commerce of the world.

It is true, in the early period referred to, the appliances for dealing with the coal were of very primitive character, and consisted of some rude coal staiths, built on the banks of the river Taff, but more primitive still was the method adopted for bringing the coal to the ships' side, viz., by pack mules and horses, in charge of drovers, carrying baskets of the precious commodity over hill and down dale to Cardiff. The construction of the Glamorganshire Canal over a century ago was, at that time, considered quite a feat of engineering skill, with its numerous locks, and marked a period of great advance in the development of the export trade, inasmuch as barges capable of carrying some 20 tons were drawn by horse power from the Merthyr and Aberdare Valleys, and thus revolutionized former methods of transit.

Bute West Dock.

Up to this period the bulk of the coal had been used by the iron masters of the district, but a market for the commodity was found in the London area, which led to the demand for Welsh Steam Coal in other places, and also led the fourth Marquis of Bute to venture practically his all on the construction of such a Dock as would lead to the cultivation of an import and export trade which would, to use his own words, "make Cardiff a second Liverpool." The Bute West Dock was opened in the year 1839. It was good enough in the year 1839, when it was first opened, and, as a matter of fact, it is good enough to-day for the class of vessel using its conveniences. Those who are not acquainted with the place would be surprised to find what extensive business is carried on in a Dock built so far back as the year named, and particularly to view the train loads of foodstuff being despatched from the extensive mills of Messrs. Spillers & Bakers to various parts of the country. Messrs. Spillers & Bakers, who own coastwise steamers, use this Dock for their large coastwise trade. It admirably suits other coastwise trades such as that carried on by Messrs. Rogers & Bright, the Dubl;n and Belfast Steamship Co., and Messrs. Gilchrists' Liverpool traders, etc. Small vessels trade between the West Dock and Ireland, France, Spain, etc., inwards with pitwood and various commodities, including large quantities of potatoes, and help to maintain Cardiff as the premier potato importing centre of the country.

Bute East Dock.

Vessels, however, increased in size, and to meet their requirements the Bute East Dock was constructed and opened in 1855 and extended in 1859. Again it may be noted that this Dock still suits a very large number of the handy-sized vessels up to 8,900 tons which frequent the Port of Cardiff. A great deal of money, however, has been spent to bring the Dock up to modern requirements. When the writer first became acquainted with the Cardiff Docks in the year 1880, all the high level coal tips in the

East Dock were worked on the counter-balance principle, that is, the wagon of coal was drawn by horse power on to the tip cradle, when a catch was released, the weight of the wagon tilted the cradle sufficiently for the coal to be shot out from the end door of the wagon into the tip shute, and from thence to the hold of the vessel. The cradle had counter-balance weights attached underneath, sufficient to bring the empty wagon back to level, after its contents had been disposed of. It was also possible to raise and lower the cradle by means of winches, operated by man power, the whole process being safeguarded by the necessary provision of brakes. A few years afterwards, however, hydraulic power was introduced, which obviated this manual labour. The tip cradles are now all operated by hydraulic rams, and the wagons are manipulated by hydraulic power capstans. Every tip in the Dock has been raised and improved, and three new tips are in course of construction. A large import trade is also carried on in the Bute East Dock.

Cardiff holds one of the premier positions for the importation of timber, and large quantities of French, Spanish and Portuguese pitwood is imported into the East Dock, as well as wines, spirits, and other commodities from Bordeaux and other Continental ports. It was in this Dock that the import trade between America and Cardiff first started, at what is known as the Atlantic Wharf, now occupied by the extensive warehouses and biscuit mills of Messrs. Spillers & Bakers. The Atlantic Liners now use the deep water Dock accommodation provided by other Docks of the system. A large flour mill is also built at the head of the East Dock, and near to a Bonded Store and other warehouses, some of which are in the occupation of the Imperial Tobacco Company, who import their tobacco leaf direct to Cardiff.

Roath Basin.

Following the opening of the East Dock, the Roath Basin, which had been a tidal harbour, was constructed in the year 1874; coal tips were erected on both sides of the

Basin, some nine in number, four of which were afterwards substituted by movable tips, the Roath Basin thus inaugurating a new method of giving despatch to shipping by employing two tips to load coal simultaneously into one vessel.

Roath Dock.

This method was further developed after the opening of the fine Dock known as the Roath Dock, in 1887, and the still larger Queen Alexandra Dock in 1907.

In the Roath Basin an appliance known as "Jumbo" had been erected. It was a crane with a portable cradle on to which the laden wagon was run, and this was transferred bodily to the hatchway of the ship, where the door catches of the wagon were knocked out, and the contents poured into the hold. The idea was good, but the process was slow, and led to the construction of what are known as the Lewis Hunter Patent Coaling Cranes, which are the coal loading appliances largely now in use at Roath and Queen Alexandra Docks. These Cranes are capable of lifting 20 to 25 tons, but instead of lifting the wagon, as did the "Jumbo" crane, pits were constructed, and a large coal box is attached to the jib crane, which is lowered into the pit. The laden wagon is worked by capstan power from the Berth Roads on to a hydraulic kick-up, which tilts the wagon, and deposits the coal into the box in the pit, after which the box is lifted by the crane, transferred to aboard ship, and lowered into the hold. An independent chain is attached to the cone-shaped bottom of the box, which, when lowered, spreads the coal evenly in the hold of the vessel with little or no breakage. Furthermore, the ship is well trimmed and there is less possibility of accident by spontaneous combustion, which may result from an accumulation of small coal on one side of the ship. These appliances are much sought after, as the coal is delivered at destination in much better condition than when shipped by means of the ordinary tip shute.

The general arrangement of these two large modern Docks for working is to confine the imports and general cargo exports to one side, leaving the other side free for the loading and exportation of coal, which obviates the confusion which would otherwise occur in dealing with mixed traffics. On the import side may be seen vessels from Europe, America, Australia, or Canada, discharging or loading into warehouses or on to the quay.

Warehouse Accommodation.

The warehouse accommodation of the whole Docks is capable of storing some 100,000 tons of goods, but even this extensive accommodation was taxed to its utmost during the period of the War.

There are also two large Cold Stores, one at Roath Dock, and one at Queen Alexandra Dock, capable of storing large quantities of frozen meat, which, together with other cold stores near to the Dock property, have a capacity of some 1,000,000 cubic feet of space.

Adjoining the Roath Dock lock extensive Cattle Lairs have been provided, together with slaughter houses and cold stores, Cardiff being the only Docks on the south side of the Bristol Channel holding a licence from the Board of Agriculture for the importation of foreign cattle. This accommodation was extensively used during the War period for imported cattle from Ireland. Possibly at this period, more than in any other in their history, Docks, Harbours, and Ports proved their value to the community.

Iron Ore Traffic.

Messrs. Guest, Keen & Nettlefolds, Ltd., rent a thousand feet of wharf space in the Roath Dock, at which they discharge their own chartered vessels conveying iron ore from the Continent to the huge Dowlais Works at Cardiff, which cover some hundred acres of ground. The iron ore traffic is worked from the wharves by means of a subway to their works, and sometimes within 48 hours the iron ore taken in is brought out in the form of pig iron or steel plates, the latter being conveyed away by rail or ship to the various shipbuilding and repairing yards of the country. Steel rails and steel sleepers, made at the Cyfarthfa works, are also loaded and despatched from this wharf to India, Asia, Africa, etc.

The Tharsis Copper Works adjoins the Dowlais Works, which is as it should be, as, after the copper is extracted from the pyrites, the residue is practically iron, and for convenience of use in the blast furnaces the purple ore, as it is called, is compressed into briquette form, passed through a furnace, and is brought out having the appearance of a solid mass of iron. As a matter of fact, there is as much, if not more, iron in these briquettes, as is contained in good spathic ore.

Imports and Exports.

Whilst Cardiff is pre-eminently a Coal Port, and as such possesses a unique position, it is also one of the principal centres for the importation and exportation of raw material and general merchandise.

In the year 1913 the imports were as follows :—

Iron Ore	746,381	tons.
Pig Iron and Ironwork	94,000	,,
Timber and Deals	166,533	,,
Pitwood	409,507	,,
Grain and Flour	311,667	,,
General Merchandise	343,212	,,
Total	2,071,300	,,

The exports, other than coal, for the same year were :—

Patent Fuel	714,730	tons.
Iron and Steel Rails and Ironwork ..	180,979	,,
General Merchandise	133,426	,,
Total	1,029,135	,,

The year 1913 has been taken for the purpose of this article, because of its being the last complete year prior to the outbreak of War.

It is gratifying that the old country, which bore the brunt of the war, seems to have recovered most quickly, and the Bute Docks, Cardiff, are getting gradually back their import and general cargo trade, as the following will show :—

Sundry Goods and Patent Fuel Imports and Exports, 1919.

Cardiff	2,132,996 tons.
Newport	551,109 „
Barry	475,251 „
Penarth	31,879 „

Dry Docks.

Space will not permit of particularising the many important industries attached to the Cardiff Docks, other than such necessary adjuncts as Dry Dock accommodation, for the repairing of vessels frequenting the Port. There are no less than twelve of these, three of which are over 600 feet in length; one with a length of 615 feet and a width of 107 feet, capable of dry docking four large steamers simultaneously.

Various Industries.

Special mention should be made of the Crown Preserved Coal Company's extensive Patent Fuel Works, which adjoin the Roath and Queen Alexandra Docks. The quantity of fuel exported from these and other works in 1913 amounted to 714,730 tons. The patent fuel briquette is manufactured from small coal, pitch, and other ingredients, compressed by hydraulic machinery, and forms a fuel practically free from dirt, which does not deteriorate either in very hot or very cold climates. Large quantities are despatched to the State-owned Railways of the Continent of Europe, South America, etc. The Crown Preserved Company are to be congratulated on the excellent canteen accommodation they have provided in these extensive works for their men.

Cardiff is a busy centre for a huge wagon repairing business, as many as twelve to fifteen thousand wagons being dealt with annually. Some of these yards also lay themselves out for the construction of new coal wagon stock. Messrs. John Williams & Son have added a new industry to their foundries, viz., steel window frame manufacturing. Quite a model workshop with excellent canteen accommodation. The output from these works represents some 6,000 square feet of window framing per week.

Extensive timber yards indicate the importance of this imported traffic, and, together with boat building yards, foundries, forges, paint, oil and tar works, coal washeries, house fuel works, malt houses, wire rope factories, and other industries, provide an object lesson to the visitor, of the resources of Cardiff in providing employment for the masses.

There are hundreds of acres of land in the vicinity of the Docks, with railway communication to all parts of the United Kingdom, suitable for further works and factories, which can be had on long lease. There is, therefore, every reason for believing that Cardiff will share in the general expansion of trade, which is bound to follow the dead period of the War.

Water Supply.

Having thus dealt with general matters, it may be added that the whole system comprises some 165 acres of water space with miles of wharves at which has been accommodated a number of the large modern liners. The entrance lock of the Queen Alexandra Dock is a thousand feet in length over all, with a width of ninety feet, and has a low water entrance, which suffices for the Roath Dock also, as both Docks are maintained at the same water level, and connected by a communication passage.

It may be gathered that to maintain the water supply of such a large acreage when it requires over 30,000,000 gallons to cover the whole dock area to the depth of one foot, care has to be taken to avoid waste. As much fresh water as possible is obtained from the River Taff, supplemented by powerful pumps on the banks of the river. This water finds its way to the lower reaches of the Glamorgan Canal, the Bute West Dock, and the Bute East Dock, as junction canals connect each of them. There is also a culvert which runs from the East Dock to the Roath Dock. The water supply, however, is augmented by three of the most powerful pumps in the world erected at the Queen Alexandra Dock entrance. These pumps are capable of throwing 100,000 gallons of water each per minute, through a culvert some sixteen feet in diameter, and the two largest Docks, namely, Roath Dock and Queen Alexandra Dock, are thus supplied.

Efforts are made to keep as much of the Bristol Channel water out of the Docks as possible, because of the large quantity of mud it holds in suspension. Some idea of this deposit may be gathered from the fact that nearly 30,000 tons of mud are dredged in and about the Docks per week.

Hydraulic Power.

Hydraulic power is maintained at a pressure of 800 lbs. to the square inch over the whole Dock system by eight power houses, some 23 engines working in harmony to maintain the pressure.

The coaling cranes, tips, and most of the numerous general cargo cranes, varying from 35 cwt. lifts to 70 ton lifts, are worked by hydraulic power.

At the Queen Alexandra Dock, however, there are eight electric and ten steam cranes, in addition to numerous hydraulic cranes. No doubt in the near future greater use will be made of electric power.

Coal Export.

The great development of the trade of the Bute Docks, Cardiff, may be gathered from the following figures :—

In 1839 the total imports and exports together amounted to	8,282 tons.
In 1860 they had jumped to	2,225,980 ,,
In 1913 (the record year) to	13,676 941 ,,

Cardiff is the premier port in the world for the shipment of coal, and first for the tonnage cleared to foreign ports.

The Customs Port of Cardiff comprises the Bute Docks, Penarth Dock, and Barry Docks.

Penarth Dock was opened in 1865 and Barry Docks in 1889.

In 1890 the coal exported to foreign countries, exclusive of bunkers, from the Port of Cardiff,

amounted to	9,481,802 tons.
In 1900	13,461,027 ,,
In 1913	19,328,833 ,,

The facilities in the Port of Cardiff are such that much larger quantities than this can readily be dealt with.

The quantity of coal exported from the whole of the Bristol Channel ports, including Newport, Barry, Cardiff, Penarth, Swansea, Port Talbot, Llanelly, etc., in 1913, amounted

to	29,875 916 tons.
of which to France and Italy were sent ..	12,811,252 ,,
and to other destinations	17,064,664 ,,

Each of the Ports of the Bristol Channel has some distinctive method of coal loading. For Cardiff the Lewis Hunter patent coaling crane system, already described, may be specially mentioned, but it might be added that all vessels at the Bute Docks are berthed straight along the quay wall, hence the reason for providing so many movable appliances. It is the appliances which are moved in the deep water Docks from hold to hold, and not the vessels. A further advantage is that two or three, or even more, appliances may be used in the loading of one vessel, and in this way 6,715 tons of coal have been loaded into a vessel in eleven hours.

At most of the other Docks of the channel the tips are erected on jetties which project from the quay wall, and, in such cases, it is the vessel which has to shift to complete the loading of the various hatchways.

PENARTH DOCK.

The outstanding berth for despatch at Penarth Dock, which is an extremely useful Dock (situated under the Penarth Head) comprises four movable tips in the Dock

Basin which can be used simultaneously into one vessel. Many coasting steamers are docked, loaded, and cleared to sea on the one tide from this berth.

Barry Docks.

Barry Docks are also splendidly equipped with high level and low level tips (fixed and movable) and with extensive siding accommodation to deal with a large coal traffic. Warehouse and cold storage accommodation is also amply provided. The Barry Docks also have a low water entrance. Amongst the industries of Barry, Rank's large flour mills are an outstanding feature.

Newport Docks.

Newport's fine modern lock and dock, claimed to be the largest in the world, has also a deep water entrance. These Docks are provided with every facility for the successful handling of the coal and general cargo trade and are also well supplied with warehouse accommodation. Steel works, tube works, and other important industries add to the prosperity of this busy centre. One feature in connection with the coal loading operations at Newport which differs from some other Docks in the Channel is that whilst at most Docks the empty wagons pass into the empty wagon sidings from the same level as the coal to the hoist, at Newport the empty wagon is disposed of from a high level gantry, the loaded wagon being raised from the quay level.

Another method of shipping coal, which should be mentioned, is the belt system in vogue at Port Talbot. Conveyors are erected at these Docks by means of which vessels may be loaded with great despatch. Baldwin's large steel works, fuel works, and other industries have settled here.

Swansea also has some very good appliances in her extensive and well equipped Dock system. The Anthracite Collieries predominate in this district. It is also the largest metallurgical centre in South Wales.

The Dock water acreage for the whole of the Coal Docks of the Bristol Channel is something between 750 and 800 acres.

The gigantic dimensions of the coal trade of South Wales and Monmouthshire may be gathered from the fact that the total output of coal in the year 1913, from some 620 Collieries, amounted to 56,830,072 tons, nearly 114,000 coal wagons being employed in working the traffic from pit to port and inland. The coalfields measure 100 miles in length by over 20 miles in their greatest width, and contain an almost inexhaustible supply of bituminous, semi-bituminous, smokeless steam, semi-anthracite and anthracite coals, which, we have only recently been reminded, is one of our greatest national assets, in exchange for the foodstuffs, raw material, and other imported commodities necessary to our very existence.

SHIPPING.

By R. O. SANDERSON, President of Cardiff Chamber of Commerce; Chairman (1916-17) of Cardiff and Bristol Channel Shipowners' Association.

Mr. Lloyd George on a memorable occasion described the shipping industry as the jugular vein of the nation.

It is literally true that South Wales as a great industrial centre depends for its very existence on the shipping industry, whilst Cardiff owes its position as the metropolis of Wales and its status as a City to its pre-eminence as a Welsh port. If Cardiff's greatness be due to the mineral wealth of Glamorgan, and particularly to the far famed Rhondda Valley, it is equally true that the development of the Welsh Coalfield and its numerous industries would never have taken place without the enterprise of the shipping community. The development of modern Cardiff really commenced when the old Glamorgan Canal was extended to the sea in 1798, but its period of rapid growth dates from the time when John, the second Marquis of Bute, risked everything and spent nearly £400,000, which he had great difficulty in raising, in constructing the present West Bute Dock, opened in 1839. At that time capital was not easy to obtain, and no one then dreamed of Cardiff's ultimate maritime greatness. The Marquis of Bute was a poor man measured by modern standards. His chief wealth lay in extensive undeveloped lands. There is no doubt that he took very great risks when he mortgaged his estates, practically to the last penny, in order to build the smallest of Cardiff's docks, and provided sea outlet for the coal and iron trades, which have changed South Wales from a land chiefly devoted to sheep farming, with a scattered population of a few thousand people, to a state of prosperity, and a million busy people within the confines of its coalfield.

It is perhaps difficult for a stranger nowadays to imagine that within the time of persons still living, Cardiff was a small township with only ten thousand inhabitants, and that the managing director of the first steamer registered at the port (the "Llandaff") is still "on 'change."

One may perhaps be permitted to mention an interesting link with the past.

There is still trading to the port a little schooner, the "Good Intent," which was built at Plymouth, out of old warship timber, in 1777—143 years ago. On the stern of this little craft are the words "The sea is His, and He made it." This vessel seems to have been well named, and to have been specially blessed by Providence, as the Captain of a few years ago stated that he had been the master for 43 years, whilst his father before him occupied that position for 52 years—a joint period of 95 years, which is probably without equal. When this vessel commenced trading to Cardiff there was no sea canal or docks, only a few river wharves—dry at low tide, and a muddy shore, and vessels approaching Cardiff took their bearings from the hay stack at Adam's Farm, Adamsdown.

Cardiff's shipping was considerably affected by the war. It is now endeavouring to regain its lost position. The record year was 1913. The trade of the first six months of 1914 was far in excess of that of any similar period; but the outbreak of war, in August of that year, caused a big decline in traffic over the remaining months. Unfortunately, since the termination of hostilities, this loss of traffic through the reduced output of coal, shortening of the hours of labour, and the lack of facilities for restoring war wastage in the equipment of the port, have prevented the trade from being regained; but it is to be hoped that the check it has received will be overcome in the not too distant future, and that new commercial records will be established.

In order to give an idea of the business of which the port is capable, a summary of the returns for the record year is appended.

	1913. Imports. Tons and Loads.	1913. Exports. Tons.	Total. Tons and Loads.
Bute Docks .. .. / Penarth Docks .. .. / Barry Docks	2,860,708	27,213,060	30,073,768

Of the exports, the coal trade represented **26,139,334** tons.

The number of vessels (steam and sail) which visited the port—including repeated voyages—and their registered tonnage, was as follows :—

	Vessels.	Registered Tonnage.
Bute Docks .. .. Penarth Docks .. .. Barry Docks	15,193	12,982,316

The total value of trade with foreign countries—imports and exports—in 1913 was over £24,000,000. Coastwise trade and the cost of bunker coals represented several millions sterling in addition, while no statistics are available regarding the value of the commerce of Cardiff inland. Having regard to the enormous increase in prices since the war began, the value of trade done, despite a decline in tonnage in 1919, was more than double that of 1913.

The volume of imports and exports in 1919 was :—

	Imports. Tons and Loads.	Exports. Tons.	Total. Tons and Loads
Bute Docks .. .. Penarth Docks .. .. Barry Docks	1,945,451	17,797,711	19,743,162

Of the exports, the coal trade represented 16,953,943 tons, a fall since 1913 of over nine million tons.

Although the aggregate trade was lower in 1919, there was an increase in the number of vessels visiting the port as the result of small craft, especially sailing vessels, being employed more largely :—

	Vessels.	Registered Tonnage.
Bute Docks .. .. Barry Docks .. .. Penarth Docks	16,592	10,458,40 3

The port is known the globe over. In addition to being the chief source of the renowned Welsh coal, upon which the British Navy and Mercantile Marine have depended, it is the principal ship repairing centre in the world, and is probably first in the ownership of cargo carrying vessels of the "tramp" class. In using the word "tramp," one must not be regarded as in any way discounting the value of the vessels owned at the port. Steamers are classed as "tramps" not because of any lack of efficiency in construction or sea-going qualities, but because they are engaged in the open carrying trade of the world and not on specific routes. Their cargoes are usually in bulk, hence the importance to them of the coalfields. Of the total United Kingdom weight of 97 million tons of exports in 1913, coals provided 73½ million tons of cargo. In 1919, coals provided 35¼ million tons.

Before the war, a fleet of several hundred steamers, with a gross registered tonnage of more than one million tons, was owned by Cardiff Firms. No port suffered more heavily through losses by enemy submarine action, but the ownership of vessels is again increasing rapidly.

There has indeed been a remarkable and even phenomenal expansion since the termination of the war. An indication of this is furnished by a reference to the list of Members of the Cardiff District Shipping Federation, Ltd. There are at present on the list no fewer than 115 companies, as compared with 57 only a year ago, and 382 vessels as compared with 213, representing, therefore, an increase of 58 companies and 169 ships. The gross registered tonnage of Cardiff owned vessels amounts to about 960,000, with a deadweight capacity of approximately 1,500,000 tons. The tramp tonnage of the United Kingdom is estimated at just over 10 million gross registered tons, so that Cardiff can claim about one-tenth of the whole tramp tonnage of the country. If anything, these figures are on the modest side, so it will be seen that the devastation wrought to shipping during the war has been much more than compensated for by the enterprise of the shipping business community of the port. Such a rapid recovery is probably unexampled in the history of any port of the Kingdom, if not of the world.

The port of Cardiff is also much used as a port of call by liners. There is no doubt that liner traffic will develop more and more, especially when the new scheme for improving railway connections between South Wales and England is extended by the construction of another bridge over the higher reaches of the Severn Estuary. The existing railway connections, which include the world-famous Severn Tunnel and the Severn Bridge—one of the most noted bridges in the Kingdom—have become totally inadequate for the great traffic of the district. The industries, commerce, and shipping facilities of South Wales might speedily be developed if traffic facilities could be extended to meet their requirements. This is recognised by the government, which has consented to take part with the Great Western Railway Company in the construction of a new bridge, which is to carry road as well as railway traffic, and may as an engineering achievement, rival the Forth Bridge.

On the other side of the Channel is Bristol, with great shipping interests of its own. It may also be of interest to recall that the "Great Britain" steamship, which was launched at Bristol on July 19th, 1843, in the presence of His Royal Highness Prince Albert, was the first iron screw steamer built for the Atlantic. She was 300 feet long, 50 feet beam, 1,000 horse power, and 3,433 tonnage. While she was under construction, Brunel, her designer, decided to substitute a screw propellor for paddle wheels, and many Royal and distinguished personages are stated to have travelled specially to Bristol to see what was then regarded as a thing of wonder. At the time of the battle of the Falklands this vessel was still in existence being used at the Islands as a coal hulk.

Notwithstanding the enormous losses entailed by the detention of vessels at the coal ports, and their failure to get cargoes outwards, shipowners have not hesitated to go on with their task, and continue to display the spirit which lifted Cardiff from a small seaside village of one thousand souls to its present proud position, and it is safe to predict that the shipping industry of Cardiff in particular and South Wales in general, in spite of every obstacle—preventable or otherwise—will continue to enhance its influence and power in the great activities of the world.

MINING FEATURES OF THE SOUTH WALES COALFIELD.

By HUGH BRAMWELL.

Physical and Geological.

The South Wales Coalfield is a typical elongated geological basin having a maximum length from east to west (Pontypool to Milford Haven) of 90 miles. The principal part of the field, from Pontypool to Llanelly (50 miles) has a fairly constant width of some 16 miles. The extension westward is narrow, being only from 2 to 5 miles across. The Coalfield is generally bounded on all sides by the outcropping and underlying Carboniferous Limestone rocks. The latter contain no seams of coal. The basin has been subjected to a heavy thrust from the south, resulting in synclinal and anticlinal folds, and their accompanying heavy gradients. From the Southern crop the measures generally dip into the basin at high angles, up to 18 and more inches per yard. Along the Northern crop the gradients are more moderate. In addition to minor folds, the field is divided into two well marked areas by a main anticlinal ridge which traverses the country for a distance of some 30 miles in an easterly and westerly direction.

The Coalfield is not overlain by recent sedimentary rocks, except to a very limited extent along the southern boundary, where Trias beds in places somewhat overlap the upturned edge of the coal measures. The main Pennant plateau comprising the greater part of the field has probably existed as a land area during the later geological periods. The surface has consequently been subjected to extensive subaerial denudation. The original plateau is now cut into narrow valleys, generally running north and south, some 500 to 600 feet in depth. The thickness of the coal bearing strata in the deepest part of the basin is estimated at 7,500 to 8,000 feet, but even there some of the original coal measure deposit has disappeared, and over a great part of the Coalfield only the Middle and Lower Measures remain. There is evidence of the denudation of the Lower Measures in the Middle Measures themselves. In the Pennant Sandstones which divide the middle from the lower series of coal seams, are several beds of conglomerate, which carry pieces or pebbles of coal which can only have come from the earlier Carboniferous rocks.*

THE COAL SEAMS.

A feature in the Coal Seams themselves is their change in character from Bituminous coal to Anthracite in well defined directions. Typical analyses are as below :—

(*From Memoirs Geol. Survey*, 1915.)

	180 Mynyddislwyn Seam (House)	125 No 3 Rhondda Seam (Manuf'turing)	259 4 Feet Seam (Steam)	13 Stanllyd Seam (Anthracite)
C Carbon ..	87·66	88·44	92·00	94·43
H Hydrogen	6·09	5·32	4·34	3·33
O Oxygen ..	4·38	5·11	2·44	1·27
N Nitrogen ..	1·87	1·13	1·22	0·97
$\frac{C}{H}$ ratio	14·39	16·62	21·19	28·36
Volatile matter	33·14	25·40	14·58	5·30
Fuel Ratio ..	2·02	2·94	5·05	17·84
Sp. Gr. ..	1·33	1·33	1·31	1·43
Ash	5·20	1·60	3·89	1·10

$$\text{Fuel Ratio } \frac{\text{Fixed Carbon}}{\text{Volatile matter}} \quad \text{Thus } \frac{100-33{\cdot}14}{33{\cdot}14} + 2{\cdot}02 \quad \text{Mynyddislwyn}$$

Horizontally in passing from east to west near Pontypool, from south-east to north-west near Pontypridd, and from south to north near Swansea, the same seams change from Bituminous to Semi-bituminous or Steam coals, and on to Anthracite, losing their volatile contents.

Vertically, a similar change is observable as between Higher and Lower seams in almost any part of the coalfield,

* Memoirs of the Geological Survey. Coals of South Wales, 1915, page 81.

it being the rule for the upper seams to be more bituminous than those below. This change may be variable in rate, but in direction it is persistent. In depth, viz., as between different seams, the rate of change is again somewhat irregular, but taking the seams in groups, it is quite persistent, and is well illustrated by the analyses of the several seams cut in a recent shaft sinking in East Glamorgan, given below :—

	Depth in Yards	Volatile Matter	
No 3 Llantwit Seam (Crops in hillside)	32	27·850%	Group A (Llantwit Group)
No 1 Rhondda Seam	462	24·760	
No 1 Rhondda Seam	462	24·760%	
No 2 Rhondda ,, ...	501	26·700%	Group B.
,, ...	524	29·000?	24¾ to 27½
No 3 Rhondda ,, ...	563	27.430%	(Rhondda Group
Hafod ,, ...	611	24·100%	
,, ...	635	25.290%	
Abergorki ,, ...	660	25·910%	
,, ...	663	25·130%	Group C.
,, ...	670	23·530%	24 to 25½
,, ...	681	25·660%	
,, ...	686	24.500%	
,, ...	718	23·200%	
2ft 9in (Upper) ,, ...	735	23·373%	Group D.
2ft 9in (lower) ,, ...	743	23.558%	20½ to 23½
4ft ,, ...	760	20·422%	(Steam Coal
6ft ,, ...	784	21·060	Group)

Along the immediate Southern Crop all the seams are highly bituminous (30 to 37 per cent. of volatile matter), but the loss of volatile matter as they dip into the Basin is rapid.

The Origin and Distribution of Anthracite in South Wales have been investigated by the Geological Survey, and the results recorded in a Memoir by Sir Aubrey Strahan and Mr. W. Pollard, published in 1915.

Mining Depths.

The fact that the Upper Seams commonly crop out on the sides of the valleys, tends to make their winning and working easy. Except, however, round the edges of the basin, the more valuable Steam coal seams lie at material depths, and mining is now practised at 500 to 800 yards below the bottom of the valleys.

Pressure.

The Middle seams particularly the Rhondda Group B, are associated with more or less massive beds of sandstone. The Lower Steam coals, Group D, are, on the other hand, chiefly interstratified with beds of comparatively soft, friable and easily weathered shales, beds of rock being thin and infrequent. This feature renders the mining of the lower seams in Wales especially difficult. An opening made in such ground at once tends to close and "squeeze" in.

The result is that as much labour and material are frequently expended in the maintenance of a Steam coal mine as are expended on the extraction and raising to the surface of the coal itself. At a Steam Coal Colliery the proportion of the several classes of men employed is as below :—

	Pre-war 8 hours winding.	
1. Getting coal	34·7%	
2. Maintaining the Mine	30·8%	
3. Traffic and Superintendence ..	18·5%	common to 1 and 2.
4. Surface work	16·0%	
	100·0%	

and recent changes have still further reduced the proportion of coal getters. Labour forms 65 to 75 per cent. of the cost of production at different mines, and the average for the Coalfield is nearer the latter than the former figure.

Pitwood Consumption.

Further, the quantity of pitwood required for the support of the roof and sides is approximately at the rate of one ton for every 30 to 40 tons of coal extracted, an amount probably three times as great as is necessary in other Coalfields. Also the quantity of debris (rubbish) to be handled underground, part of which is sent to the surface, may in some cases equal the quantity of coal raised.

"Slips."

The Coal produced in South Wales has a characteristic appearance of its own. The usual cubical fracture of coal is partly developed in the Upper Bituminous seams (Llantwit series and their equivalents) but the coal in the Lower seams is divided by " slips " or joints generally 1 to 2 feet apart, lying at an angle of about 45° between the roof and floor. Their direction is remarkably constant over large areas, and is more or less the same as that of a well-defined system of Main Faults. For instance :—

	Direction of " slips "	Direction of Main Fault System
At Aberdare	N. 44 W.	N. 44 W.
At Pontypridd	N. 40 W.	N. 37 W.
At Maesteg	N. 43 W.	N. 43 W.

The presence of these " slips " further renders coal cutting by machinery (undercutting) inapplicable and unnecessary.

Examples of " Cone in Cone " structure in the coal itself are to be found in certain seams and districts.

The result is that the Welsh Steam Coal is produced in large and irregularly shaped lumps, which again break with an irregular fracture, and produce a fine dust in the form of " flakes " as if caused by " attrition " rather than by direct " crushing."

Mining.

Such conditions have led to the adoption of systems of mining and types of plant, differing from those developed in other coalfields.

The size of the lumps prevents the use of " hopper-bottom " wagons on the railways, all the trucks have end doors and are discharged by end " tipping." The capacity of the vehicle used for transport underground is materially larger than that commonly used elsewhere. The Welsh " tram " carries anything from 1 to 2 tons of coal. The " tub " of other coalfields carries 8 to 15 cwts. Props or posts set for support underground at right angles to the roof and floor, are only adopted in the working faces, the roadways are protected by special systems of timbering developed to meet the "squeeze" from both roof and sides. Arching of the roads in brick, stone, and with curved iron girders, is very largely practised, and concrete lining is being introduced. The mechanical equipments having to deal with a large unit in the "tram" have to be on a proportionate scale. The "pit pony" of 12 and 13 hands is almost unknown, the Welsh "pit horse" being a short legged miniature dray horse of 14-2 and 15 hands.

WINDING OR HOISTING.

Average hoisting or winding plants have capacities of 100 to 200 tons per hour, larger equipments running up to 350 tons per hour. The actual running time on coal load throughout the year is, however, limited by traffic requirements other than coal, customary shifts, holidays, etc., so that an equipment for 350 tons per hour would have a maximum annual capacity of 650,000 tons of coal or thereabouts, or roughly a 25 per cent. productive load factor.

POWER.

On the surface the chief motive power is "steam," but there are few collieries now without their auxiliary electric plant, and in some instances steam has entirely been replaced by an electrical supply. Central electrical stations for groups of collieries are becoming common, and the South Wales Power Company provides a public supply over a large part of the Coalfield for those who prefer to buy, rather than to generate for themselves.

Underground the power almost universally employed is compressed air, transmitted from the surface.

Depending upon the mining conditions, electricity has also been largely introduced underground, but in the Steam Coal Mines this is generally confined to the neighbourhood of the shaft bottoms.

Shafts.

The early shafts in South Wales were frequently oval or eliptical in shape, but all recent ones are circular. The size of later shafts may be 20, 22, and even 25 feet in finished diameter.

Ventilation.

Except in the neighbourhood of the outcrops the mines are generally "gaseous." Outbursts of gas, whilst not unknown, are not common, but large volumes of gas are persistently given off, especially in virgin areas of development. The gas does not usually issue from the coal itself, but is chiefly yielded by small "blowers" from the roof and floor. Ventilation is almost entirely produced by Centrifugal Exhaust Fans. A mine producing, say, 1,000 tons of coal a day, will have a ventilating current of 250,000 cubic feet of air per minute, viz., about 360,000 cubic feet of air per ton of coal extracted, on a 24 hour basis.

Extraction of the Coal.

Complete extraction of the seam in one operation, viz., Longwall work, is the almost universal practice, systems of pillar and stall work being now of exceptional application.

Complete extraction in one operation offers the choice between two distinct methods of dealing with the exhausted space. The object in each is the same, viz., to obtain the tight closing up of this space as rapidly as possible.

Complete Stowing.

Where, as in South Wales, the associated strata are comparatively soft and bend readily, coupled with the fact that the coal seams generally yield in working considerable quantities of rubbish, the system of complete stowing accompanied by the bending and settling down of the roof without breaking has naturally developed. In other coalfields where the strata are harder and tend to break rather than to bend, complete stowing is not common. The roof is, on the contrary, intentionally given every facility to break and come down in instalments behind the working faces, thus closing up the exhausted space as early as possible without violent subsidence.

Loss in Working.

Complete stowing in Wales entails a loss of "pitwood," and has been one of the factors leading to the loss of part of the small coal produced in working. At first sight this appears to be an easily avoidable loss. The system, however, results in very complete *all round* extraction, so much so that the Royal Commission on Coal Supplies (1903) found that the "gross loss in working" in Wales did not exceed that in other coalfields.

Differences in the character of the seams now generally worked as compared with those first operated upon, are leading to a material reduction in this loss of coal. Recent evidence points to a loss from this cause at those mines where such loss occurs, of 5 to 7½ per cent. of the calculated contents of the seam, and the number of seams operated upon where no such loss occurs is increasing. It is probable that this change will in some seams entail the abandonment of complete stowing, and the adoption of the system of breaking the roof behind the face.

Trade.

The character of the coal and the position of the coalfield in relation to the seaboard have led to South Wales becoming primarily an exporting district.

A Historical and Statistical note is included in the handbook. (See page 169.)

The several areas of the Coalfield from which the different classes of coal (based on Carbon and Hydrogen contents), are drawn, are approximately shown on the map which also shows the position of the principal towns, ports and railways.

BRIQUETTES.

The utilization of small coal for the manufacture of Patent Fuel or Briquettes has probably reached greater proportions in Wales than in any other Coalfield, and a special note as to this industry will be found in the handbook. (See page 188.)

COKING.

So far Metallurgical Coke has been the " product " aimed at, and material progress has been made in recent years in this industry in the Coalfield. Welsh Coal, however, is low in volatile matter and requires a high temperature to coke, and these two features result in low yield of tar, ammonia, and other bye-products. The industry is thus on a somewhat different commercial basis to that of other coalfields. A plant coking Welsh coal with 20 per cent. volatile matter has in practice a tar yield of 52 lbs. per ton of coal coked, and a sulphate of ammonia yield of 18 lbs., compared with such figures as 88 lbs and 28lbs. respectively for another coalfield, where more bituminous and more easily coked coal is available.

For these reasons it would also appear that Low Temperature Distillation, having the Tar, Tar Oils, Ammonia, etc., as the chief products, and coke the bye-product, will be introduced into other coalfields before it comes to Wales, unless such works are combined with special mining operations in the more bituminous coal of the Southern Crop.

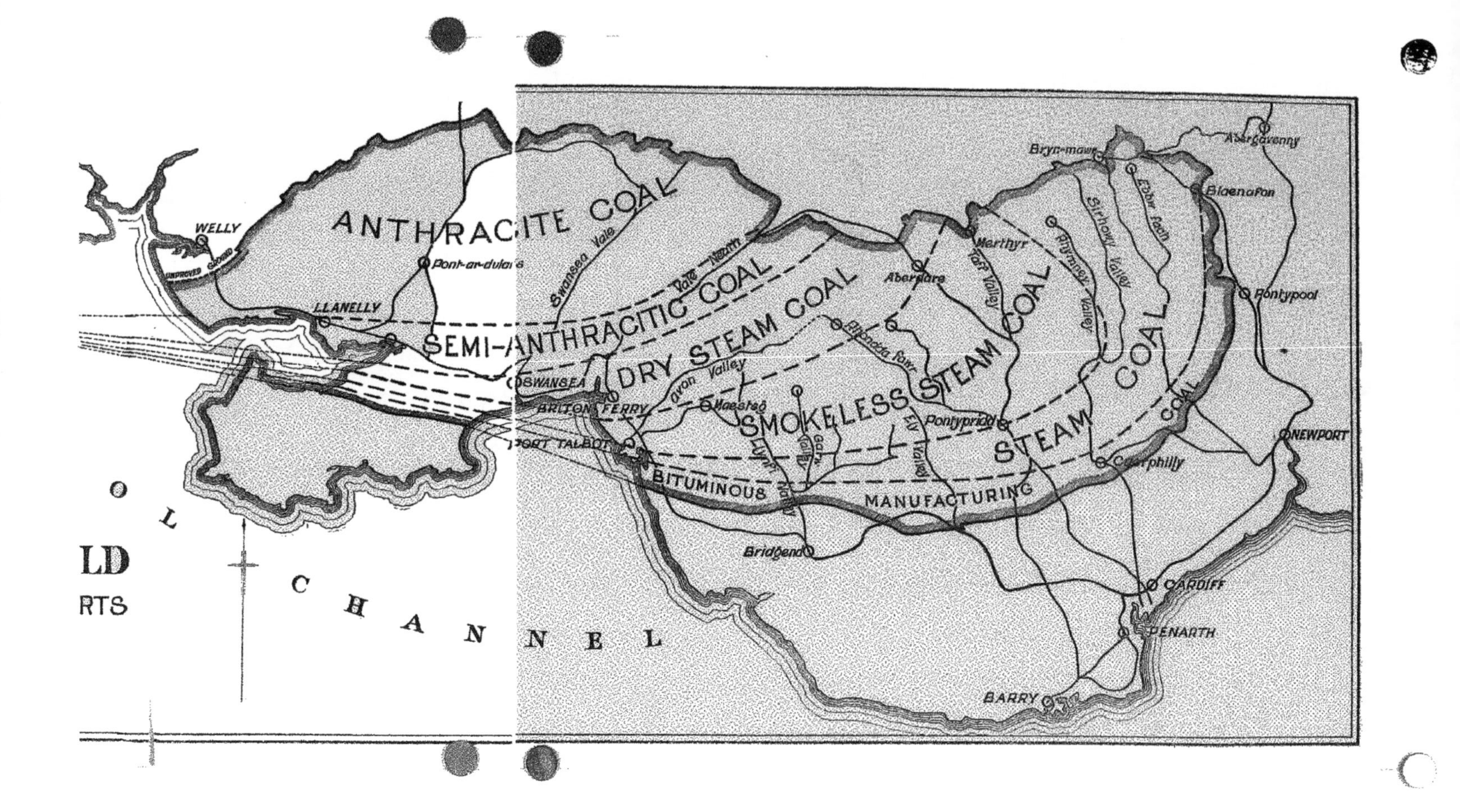
ANTHRACITE COAL
SEMI-ANTHRACITIC COAL
DRY STEAM COAL
SMOKELESS STEAM COAL
STEAM COAL
BITUMINOUS
MANUFACTURING
WELLY
IMPROVED GROUND
Pont-ar-dulais
LLANELLY
SWANSEA
BRITON FERRY
PORT TALBOT
Swansea Vale
Vale Neath
Avon Valley
Maesteg
Llynfi Valley
Garw Valley
Ely Valley
Rhondda Fawr
Aberdare
Merthyr
Taff Valley
Pontypridd
Rhymney Valley
Sirhowy Valley
Ebbw fach
Bryn-mawr
Abergavenny
Blaenafon
Pontypool
NEWPORT
Caerphilly
Bridgend
CARDIFF
PENARTH
BARRY
CHANNEL
LD
RTS

COAL TRADE OF SOUTH WALES.

By FINLAY A. GIBSON.

THE South Wales mineral basin extends from St. Bride's Bay in Pembrokeshire on the west to Abersychan in Monmouthshire on the east, embracing a portion of Pembrokeshire, a large portion of Carmarthenshire, nearly the whole of Glamorganshire and Monmouthshire, as well as a portion of Breconshire, about 89 miles from east to west, and varying in breadth from north to south at different points from 21 miles at its greatest breadth to about 1½ miles, and extending over about 1,000 square miles. The coal measures are divided into three series, viz., The Upper Pennant, the Lower Pennant, and the Lower or White Ash series, which contains the principal seams of coal. The basin is traversed for many miles in the eastern portion by an important anticline which converts that portion of the field into two troughs, while in addition to this main anticline there are numerous local anticlines of minor importance, and also large faults. The coal measures vary in quality from highly bituminous on the east to pure anthracite in Carmarthenshire and Pembrokeshire. According to the late Lord Merthyr there are in Pembrokeshire 10 seams of an aggregate thickness of only 28ft. 8in.; in Carmarthenshire there are, at one point, 18 seams on the north crop yielding 47ft. of workable coal, while on the south crop there are 34 seams yielding 83ft. of coal; in Glamorganshire there are 48 seams yielding 124ft. 6in. of coal, 22 yielding 91ft. 9in., 23 84ft. 10in, 22 94ft. 2in., 14 65ft. 11in., 26 66ft. 8in., and 26 70ft. 11in.; while in Monmouthshire there are 21 seams yielding 47ft. 2in. on the north crop, 15 seams on the south crop yielding 46ft. 10in., and 11 seams on the eastern outcrop yielding 38ft. of coal.

In his estimate of the available resources in seams of 12in. thick and upwards, the late Lord Merthyr gave the following approximate figures of net tonnage :—

District.	Net Tonnage.
Monmouthshire	2,743,508,074
Glamorgan (East)	11,070,360,468
Glamorgan (West) (Carmarthen and part of Brecon)	14,349,335,795
Pembrokeshire	172,583,813
Total	28,335,788,150

Allowing 5 per cent. for consumption by the collieries themselves in the raising of coal, Lord Merthyr arrived at a net marketable quantity of 26,918,998,743 tons. These estimates were given in 1904. Since then approximately 800 million tons of coal have been extracted, leaving 26,118,998,743 as the present approximate quantity of marketable coal according to the estimates of Lord Merthyr. The 1904 estimate was classified as follows :—

Quality.	Tons.	
Bituminous	8,618,688,965=30.42%	
Anthracite	6,310,292,214=22.27%	
Steam (Western Division)	4,076,424,971	47.31%
Semi-bituminous or second class steam	5,393,724,590	
First class steam	3,936,657,410	
Total	28,335,788,150	

It is from the collieries supplying the first class steam coal that the Admiralty and the mercantile marines of the United Kingdom and foreign countries have met their increasing requirements during the past fifty years, and as it is upon its steam coal resources that South Wales has made its reputation as the leading coal exporting centre in the world, it is of interest to note that in 1904 the late Lord Merthyr estimated the quantity of best Welsh steam coal remaining unworked in certain best steam coal seams of one foot thick and upwards in collieries on the Admiralty list at 3,240,182,734 tons, in other collieries not on the Admiralty list, 174,793,547 tons, and in unlet areas 521,681,129 tons.

Output.

The chief methods of working adopted in the development of the coalfield have been those of the long wall and pillar and stall systems, but in the new collieries the long wall system has been more generally adopted. During the first half of the century the coal mines were regarded primarily as the ancillary undertakings of the iron works, but in the seventies the coal trade had asserted not only its independence, but its supremacy, over the South Wales iron industry, and the great stimulating factor in the remarkable development which subsequently took place was the substitution of steam for sail in ocean navigation. When the first Association of Colliery Proprietors was formed in 1864 under the title of the Aberdare Steam Coal Collieries Association, it consisted of only 11 firms, producing an output of 1,600,000 tons. The majority of the coal masters refused at that time to become associated. In 1854 the output of the whole coalfield was only 8½ million tons. By 1860 it had increased to 10¼ million tons, and the outputs at decennial intervals since 1860 have been as follows :—

Year.	Output.
1860	10,255,563
1870	13,594,064
1880	21,165,580
1890	29,415,025
1900	39,328,209
1910	48,699,982
1913	56,830,072

During the whole of this period the industry was entirely under private enterprise. At the outbreak of the war the Government took control of the output of the collieries producing Admiralty steam coal, and in December, 1916, all the coal mines in South Wales passed into the possession of the Board of Trade. The outputs since 1913 have been as follows :—

Year	Output
1914	53,879,728
1915	50,452,600
1916	52,080,709
1917	48,507,902
1918	46,716,535
1919	47,522,306

In July, 1909, the Eight Hours Act was brought into operation in the coalfield, and in July, 1919, the Seven Hours Act. How great the effect on employment has been the development of the mineral resources may be judged from the fact that whereas in 1888 the number of persons employed was 91,423, it is at present over 260,000—a nearly three-fold increase.

Exports.

The impetus for this development has come mainly from the foreign and bunker demand. Up to the end of the 18th century, the woolsack was the emblem of Britain's economic strength. To-day that emblem is more truly represented by a coalsack. Before the discovery of the steam engine we were primarily an agricultural country, but with the discovery and improvement of the value of steam as the motive power of industry there began that revolution which in the course of a century was to transform these islands from a farm into a workshop, and we were able to do this because our possession of coal in abundance and at easily worked depths, provided us with a commodity which found a ready market abroad, and which through the process of exchange gave us command over the raw materials and the grain of foreign countries. Cheap coal cargoes provided outward bound tramp steamers with an ideal cargo, and this outward freight enabled shipping to return with ores, grain, timber and other bulky commodities at a much lower cost than would otherwise have been possible.

The growth of the export trade of South Wales is shown in the following table of cargo and bunker exports :—

Year	Cargo Exports Tons	Bunkers Tons
1870	....	....
1880	6,893,839	....
1890	12,507,636	1,730,000
1900	18,457,238	2,667,000
1910	25,215,303	4,340,714
1913	29,784,930	4,993,728

Under this trade expansion the vessels cleared from the Bute Docks increased from 1,890,230 tons register in 1874, to 6,167,933 tons in 1913. To this great and indispensable trade to the United Kingdom as an industrial centre the Great War of 1914-18 gave a shattering blow. While the war was in progress exports were determined by the exigencies of maritime warfare, and since the termination of the war the Government has imposed restraints on exports in order to safeguard a sufficiency of supply for home industries. The cargo and bunker exports since 1913 from the Bristol Channel ports have been as follows :—

Year	Cargo Exports Tons	Bunkers Tons
1914	24,475,551	4,423,152
1915	18,601,896	3,723,658
1916	17,417,707	3,579,223
1917	19,893,015	3,117,914
1918	17,000,834	2,487,110
1919	20,229,802	3,502,453

Under the control scheme at present in force the foreign cargo export trade of South Wales is further reduced to a quantity at the rate of 13½ million tons per annum (excluding coal for the Admiralty, coke and patent fuel), or less then half the pre-war volume of that trade.

It is estimated that at the present time the annual wages bill amounts approximately to between 55 and 65 millions sterling.

WAGES.

Prior to 1875 the regulation of wages was largely based on a system of individual bargaining. There was little organisation either on the side of the employers or on the side of the workmen, and disputes were frequent. Thus there were no less than three important strikes in 1871, 1873, and 1875, and it was the result of the latter that the first attempt was made to organise the system of wage regulation. In that year the first sliding scale was adopted. Its guiding principle was the addition to the base day wage or piece rates of a definite percentage automatically determined by the actual selling price of coal. The Joint Committee created under the Agreement of 1875 dealt with questions other than wages, and the sliding scale system remained in operation until 1903, when it was displaced by the Conciliation Board system. Under the latter system the principle of regulating percentages according to fluctuations in periodically ascertained average selling prices was maintained, but other factors, such as that of the volume of trade were allowed to be brought into consideration, while the deciding vote on any application for a revision of wages rates was left in the hands of an independent Chairman. It was the advent of the Independent Chairman that marked the great change from the previous system, for under the Conciliation Board the old automatic relations between prices and wages ended.

During this extended and systematic application of the principle of collective bargaining both the miners and the employers organisations increased in numbers and in influence, and at the present time the South Wales Miners' Federation claims a membership of 185,044, while the Monmouthshire and South Wales Coalowners' Association consists of the owners of 336 collieries with an assured tonnage for the current year of 45,000,000 tons. Sir J. W. Beynon is the Chairman of the Association and the Hon. W. Brace, M.P., the President of the South Wales Miners' Federation.

In the following table a record is given of the average selling price of large coal and the average general wage rate per annum since 1880, as for the purposes of the sliding agreement of that year the rates prevailing in 1879 were accepted as a basis.

Year ended December,	Average f.o.b. price. Large Coal.	General Wage Rate on 1879 standard.
1880	8/5·51	4·58%
1890	13/0·30	43·64%
1900	15/1·86	53·23%
1910	14/9·55	49·68%
1913	16/4·82	58·95%
1914	17/5·99	60·00%
1915	20/10·83	78·05%
1916	24/5·19	103·75%
1917	26/11·59	133·75% plus 1/6 per day.
1918	30/9·19	133·75% plus 3/- per day.
1919	41/0·65	133·75% plus 5/- per day.

The control of the Conciliation Board over wages practically ceased when the Government took over the control of the mines in December, 1916, while thenceforward the power of the Conciliation Board in the settlement of disputes on other questions diminished and that of the Coal Controller increased. In March, 1918, however, a Joint Disputes Committee was formed with the object of removing the complaints of miners against dilatoriness in the operation of the machinery of the Conciliation Board, this Committee during the past two years has rendered much useful work under the Chairmanship of Mr. Evan Williams, who, in addition to being the permanent Chairman of the Coal-owners' side of the South Wales Conciliation Board, is this year, for the second successive term, the President of the Mining Association of Great Britain.

IRON AND STEEL.

By DAVID E. ROBERTS.

A hundred years ago this district was probably the most important iron-making centre in the world.

Iron-making has, of course, been known from very ancient times, and in the olden days the fuel employed was charcoal. Remains of smelting grounds have been discovered in various parts of this country, but the progress was never great, partly because of the excessive inroads on the forests—something like one-third of an acre of woodland being necessary to make a ton of iron. The old smelting positions were not necessarily at the ore deposits, but at some point convenient to both the ore and the charcoal—both supplies being carried on the backs of mules to the smelting furnaces.

The actual date when it was discovered that smelting of ores could be done by coal is not quite certain. The process was known about the year 1620, at which date a Worcestershire ironmaster named Dudley was granted a patent for the process. It was, however, quite a hundred years before the process became finally established.

About the year 1750 rapid developments commenced in this neighbourhood. It was soon discovered that along the edge of this coalfield, particularly the northern edge, were to be conveniently found, underlying one another and close to the surface, ore, coal, limestone, fireclay, in fact, all the materials necessary to the production of iron ; and in the course of a few years various works were established at numerous points along this line of outcrop, and altogether well over a hundred blast furnaces have been built in the neighbourhood.

All these old works prospered, and it is quite safe to say that at the very commencement of last century this district was producing a larger quantity of iron than any other district in the whole world. This process went on uninterruptedly for a considerable time, when two factors arose, each of which had important bearings upon the local trade. In the first place, the importation of foreign ore, which on account of its greater richness and purity, was very much more profitable to work, undermined considerably the previous special advantages of the inland works at the outcrop. Then again, the introduction of the Bessemer process, and the general change over from finished iron to finished steel, involved large expenditure on the remodelling of the various works, so that to-day, of these works along the crop, only three survive on the old position—namely, Dowlais, Ebbw Vale, and Blaenavon.

As indicating the difficulties frequently met with in making any process a success, it is interesting to note that Sir Henry Bessemer brought his invention first to public notice at the Cheltenham Meeting of this, the British Association, in 1856; and although the Dowlais Works took up the right of manufacture at once, yet it was about twelve years later before they had succeeded in establishing the regular manufacture of steel rails as a commercial success.

Following on the developments of the Bessemer process came at a later date the Open Hearth System, as devised by Siemens and Martin; and an early example of manufacture under this process in the district was at Landore, where Siemens worked some melting furnaces,—these being subsequently taken over by Messrs. Wright & Butler, the former, now Sir John Roper Wright, Bart., being head of the firm of Messrs. Baldwins, who have carried out considerable developments near the site. Later still was discovered the Basic Process, which is applicable to either the Siemens Open Hearth Furnace or the Bessemer Convertor, and consists in the employment of a special lining to the furnaces, which itself has an important bearing upon the manufacture. This process was the invention of Thomas and Gilchrist, the former a London clerk, being the genius and real inventor, and Gilchrist (his cousin, who gave him valuable assistance) being a chemist at Cwmavon Works, near Port Talbot. Until the introduction of this system, no ore beds containing phosphorus could be satisfactorily used. Under the invention, however, due to the special lining employed, the phosphorus is readily extracted from the iron in the process. A basic slag is also produced which, after treatment, is converted into a very serviceable fertilizer.

It is interesting to note that, at the time of the Franco-Prussian War, when France gave up the territories of Alsace and Lorraine, it was well-known that these countries held very large deposits of ore. Up to that time, however, these beds had not been used, because of their phosphoric character. The Germans took over this territory in 1871, and this Basic discovery was made about 1872, which immediately rendered the whole of this valuable deposit useful; and the Germans have built many magnificent works on the site. This discovery, therefore, may almost be said to form the foundation of the modern German steel trade.

Of recent years, local interests have rather favoured the establishment of works nearer to the coast—thereby saving, as against inland situations, the carriage of ore from the docks to the works, and similarly the carriage of manufactured material from the works to the docks for export. Examples of such improvements are the extensive Cardiff Dowlais Works in this City, belonging to Messrs. Guest, Keen & Nettlefolds, and the modern works at Port Talbot, belonging to Messrs. Baldwins.

The old furnaces referred to earlier, though numerous, were very small, when viewed from our present standpoint. From an old furnace charge-book for 1800 it is found that the output per furnace varied between 20 and 30 tons a week, so that a hundred of them would only turn out say an average of 2,500 tons per week. To-day there are fewer furnaces—probably the number in blast in the district at the same time is well under 20. Yet at Cardiff, Dowlais, and Ebbw Vale there are (and at Port Talbot there soon will be) furnaces that will *singly* turn out nearly as much per week as the average figures mentioned above for all the old furnaces put together. It will therefore be seen that, although the works in the neighbourhood are now comparatively few in number, yet they are very large units and jointly they produce an output approaching towards a million tons of steel per annum.

A modern steel works consists of blast furnaces, steel melting furnaces, and rolling mills, and a few brief particulars follow.

Blast Furnaces.

A modern furnace is about 80 or 85 feet high, with a hearth diameter of 12 or 14 feet, and requires from 30,000 to 35,000 cubic feet of air per minute, compressed to about 15 lbs per square inch, and heated to a high temperature.

To heat this air a set of four or five stoves is required for each furnace—and these are about 90 feet high, and say 21 or 22 feet in diameter, and are filled with chequer brick work. This is a honeycomb arrangement heated up by the waste gases from the furnace—the heated interior subsequently warming the air blast for the furnace.

Suitable equipment for handling and storing the materials and depositing it at the top of the furnace is also necessary, and likewise blowing plant for providing the air supply, which may be either by steam turbine or gas engines—each system having advantages peculiar to it.

The pig iron produced by the blast furnaces may be either cast into moulds and then broken up for remelting, or it may be taken direct in a ladle to the steelworks for further treatment. It may be interesting to note here that the system (now quite universal) of carrying molten iron considerable distances in ladle carriages, was first practised at Ebbw Vale nearly fifty years ago. It was a very bold step to take in those days.

Steel Works.

The two principal processes of making commercial steel are the open hearth (*i.e.*, the Siemens-Martin process) and the Bessemer process. Both can be worked on either acid or basic linings.

The open hearth furnace is a strongly made rectangular brick-lined structure, containing a shallow bath. A standard

60-ton furnace would have a bath of metal, say 35 feet long, 14 feet wide, and, say 18 inches deep. At either end of the furnace, ports are arranged of suitable size and proper inclination to admit the heated gas and air which sweep along the surface of the bath, thus carrying on the process of the conversion into steel. The furnace is reverberatory and is reversed at regular intervals, so that the large volumes of chequerwork provided are alternately being heated on the one hand by the waste gases, and on the other, are giving up their accumulated heat to the air and gas, on their way to the furnaces for combustion and use.

These furnaces are to-day charged mechanically from the working platform through suitably placed doors in the furnace structure, and the charge consists of pig iron, either molten or cold, steel scrap, ore, lime, etc. These open hearth furnaces are sometimes made to tilt. In the fixed type above described, the finished steel is emptied by a tap-hole near the bottom of the bath, which has to be pierced and opened by hand labour each time the furnace is tapped. In the tilting type of furnace, the furnace structure is carried on a pair of rockers, so that whenever it is desired to draw off the finished material, the whole furnace is mechanically rolled forward, and so the pouring takes place. There is a further modification of this latter furnace wherein a much larger quantity of material is carried, say 200 tons. Here the process is what is called continuous, *i.e.*, only a part —say two-thirds of the finished heat—is run off the furnace, the remainder being left in to mix with the incoming fresh charge, and to help its conversion.

Each type has its own special advantages. In the Bessemer process a large jar-shaped vessel is used to hold the metal, which is hung on trunnions, so that it can be turned over when necessary. The vessel or jar is heavily lined with bricks, and has a double bottom or blast box, perforated on the face next to the metal. Into the blast box air under high pressure is blown, and this passes through the numerous perforations, and is then through the molten metal itself, thus carrying on the process of conversion.

When the heat has been fully blown, the vessel turns over and pours the steel into a suitable ladle for teeming.

In all these processes the molten steel is teemed out of the receiving ladle into ingot moulds of a size and shape suitable to the mill they supply. Subsequently the mould is stripped from the ingot, and the latter, after reheating or soaking is ready to be rolled in the mill.

Each type of steel process and furnace just referred to is to be found in this district, and the ingots produced for rolling vary in size and weight from about 10 tons down to one ton.

Mills.

There are in the neighbourhood many systems of rolling and types of mills, which to describe even briefly would require considerable space.

It may be stated, however, that plates suitable for boiler-making and shipbuilding of every standard size and thickness are produced.

Also large quantities of rails and structural material, as well as sheets and rods, and lastly, tyres.

The author would apologize for the elementary character of this sketch. It was drafted to give a mere outline to the numerous members of the Association who are possibly not familiar with the methods of this particular industry.

SHIP REPAIRING.

By T. ALLAN JOHNSON, J.P., M.I.N.A.

Bristol Channel Dry Docks and Ship-Repairing Facilities.

There is probably no district in Great Britain which has shown such progress in the provision of ship-repairing facilities as South Wales and Monmouthshire.

The rapid growth of the ports of Cardiff, Barry, Penarth, Newport, Swansea, and Port Talbot, due to the ever-increasing demands for Welsh Coal, led to a vast amount of tonnage being dealt with, which naturally necessitated dry dock facilities for ship-repairing of a high order.

The district has made ship-repairing a speciality, and individual firms have vied with each other in the endeavour to secure despatch and efficiency.

Shipowners are now assured that in sending their vessels to the Bristol Channel district, the necessary repairs will be accomplished with efficiency, at low competitive prices, and in the minimum of time.

Much has been necessary to build up this reputation. It has only been accomplished by the introduction of the very latest plant, machinery and labour-saving devices, the retention of highly skilled workmen, and the unremitting attention of the management to the work in hand. Up to the present time the district, by reason of the War, has found itself engulfed in work, a large proportion of which in normal times would have been done on the Continent, or in the North-east Coast ports.

During the War, Antwerp, Rotterdam, and other ship-repairing centres were out of the running, and with the North-east Coast largely engaged on constructive work, vessels requiring repairs have been diverted to the West Coast ports—South Wales receiving the bulk of the business.

A complete historical survey would take considerably too much space, and therefore it is only possible to refer to a few instances of the remarkable progress and development of the ship-repairing industry of the Bristol Channel area.

A nation which aspires to the possession of a big Mercantile Marine must, in addition to large shipbuilding facilities, possess equally extensive facilities for ship-repairs.

The United Kingdom is in the fortunate position of both.

Ship and engine repair plants have quite kept pace with the extraordinary development of our Mercantile Marine. They have, in conjunction with the dry docks, which are a necessity for such work, rendered possible the growth and efficiency of the Merchant fleets which operate under the British Flag.

Every port in the Bristol Channel has, during recent years, enormously increased its repairing plants so as to cater at hand for the wants of the shipowners sending their ships there.

With regard to the Bristol Channel factor in British ship-repairing, its growth has been phenomenal, and, perhaps in no other centre is the work of such a varied and cosmopolitan character. This is due to the diversified nature of the tonnage which either bunkers or loads at the ports. Tramps and liners of every nationality frequent the Channel, and wherever shipping congregates, facilities for repairs and overhauls are always in demand.

It would be altogether impossible to estimate, with any degree of exactitude, the capacity of ship-repairing plants in the Channel, especially as they are of such elastic character that they can, with no great effort adapt themselves to any emergency that may arise.

An adequate notion of what our ship-repairing establishments can accomplish when the necessity arises, is furnished by their great War record. During that trying period, the British Mercantile Marine would never have been able to have kept the seas as it did, had it not been for the facilities available for effecting repairs to ships damaged by enemy action, or by collision, or stranding.

The Bristol Channel, not many years ago, with all its possibilities, had but a very small accommodation for ship-repairs. With the advent of steam to replace the sailing ship, the trade of the Channel materially increased, and within the past 25 years, this district has unquestionably attained the proud position of being the premier area in the world for ship-repairing.

The fact that vessels from all parts of the world are sent here for repairs is ample evidence that the establishments undertaking such work at these ports have appliances and accommodation unrivalled by any other centre to undertake all classes and magnitude of repairs.

The following is a list of the names and dimensions of the principal Drydocks, Pontoons, and Slipways in the Bristol Channel :—

BARRY.

Name.	Length.	Breadth.	Owners.
	feet.	feet.	
Private Dock ..	784.6	55	Barry Graving Dock and Engineering Co., Ltd.
,, ..	625.0	65	
Public Dock ..	867.6	54	Barry Railway Co.

CARDIFF.

Name.	Length.	Breadth.	Owners.
	feet.	feet.	
Private Dock ..	440.0	52.3	Mount Stuart Dry Docks Co., Ltd.
,, ..	420.0	52.0	
,, ..	550.0	66.0	
,, ..	419.0	60.0	Junction Dry Dock and Engineering Co., Ltd.
,, ..	600.0	55.0	Bute Dry Dock and Engineering Co., Ltd.
Pontoon Dock ..	360.0	..	Mercantile Pontoon Co., Ltd.
Private Dock ..	408.0	48.5	Hills' Dry Docks and Engineering Co., Ltd.
,, ..	400.0	45.0	
,, ..	235.0	38.0	
,, ..	635.0	62.6	Cardiff Channel Dry Docks and Engineering Co., Ltd.
Pontoon Dock ..	335.0	..	
Public Dock ..	600.0	60.0	Cardiff Railway Co.
Slipway ..	900.0	340.0	Messrs. Elliott & Jeffery.
Gridiron ..	450.0	..	

NEWPORT.

Name.	Length.	Breadth.	Owners.
	feet.	feet.	
Public Dock ..	523.0	50.0	Alexandra (Newport) Docks.
Private Dock ..	375.0	57.6	Mordey, Carney & Co., Ltd.
,, ..	350.0	47.6	
,, ..	289.0	42.6	
,, ..	222.0	36.0	
,, ..	712.0	65.0	Tredegar Dry Dock Co., Ltd.
,, ..	415.0	62.0	Messrs. C. H. Bailey.

PENARTH.

Name.	Length.	Breadth.	Owners.
	feet.	feet.	
Pontoon Dock and Slipway	380.0	..	Penarth Pontoon Co., Ltd.

PORT TALBOT.

Name.	Length.	Breadth.	Owners.
	feet.	feet.	
Private Dock ..	422.0	60.0	Port Talbot Graving Dock and Shipbuilding Co., Ltd.

SWANSEA.

Name.	Length.	Breadth	Owners.
	feet.	feet.	
Private Dock ..	480.0	42.6	Ocean Dry Docks Co., Ltd.
,, ..	375.0	47.0	
,, ..	370.0	60.0	Harris Bros., Ltd.
Gridiron ..	300.0	..	
Private Dock ..	455.0	60.0	Prince of Wales Dry Dock Co., Ltd.

In addition to the foregoing, private Dry Dock owners, there are also several ship-repairing firms established in Cardiff, Barry, Newport, and other Ports who utilize the public Dry Docks, these firms include branch works of several of the private Dry Dock owners, also Messrs. T. Diamond & Co., John Shearman & Co., John Rogers & Co., Barrett & Co., Hodges & Co., G. W. Thompson, &c.

The following is an approximate table of wages paid :—

	£	s.	d.
Year ending 30th June, 1915	870,094	2	6
,, ,, 1916	1,319,575	7	10
,, 1917	1,871,665	15	9
,, 1918	2,511,670	14	1
,, ,, 1919	2,676,139	2	0

The approximate number of men and boys continuously employed at the various establishments would be somewhere about 15,000.

Much secrecy has been maintained in connection with the work carried out in ship-repairing yards during the War, and it is no exaggeration to state that the ship-repairers of the Bristol Channel have undertaken over 75 per cent. of the whole repairing work, and saved the country some millions of pounds by repairing and making good vessels that suffered damage of such extensive nature that before the War they would have been broken up for scrap.

In addition to repair work to hulls and engines, our repairing forces were utilised in other directions. Ships had to be converted to adapt them for uses other than they had served during peace times. Troopships and Hospital ships had to be fitted and repaired ; emergency arrangements had to be made to enable ships of ordinary type to do the work of oil-tankers.

Many hundreds of merchant steamers had to be equipped with guns for defensive armament, also various devices to protect them from mines and submarines, and it is to be noted that 75 per cent. of all vessels so equipped were fitted out by the Bristol Channel ship-repairers.

In conclusion, it is evident that the Bristol Channel engineers and ship-repairers have probably seen more of the disastrous effects of the War than any other body of workers in the whole kingdom.

It has been their duty to restore to a sea-worthy condition vessels damaged by German piracy, and others which have suffered under the heavy strain of War conditions. It is therefore, not too much to say that these men, who had to control and direct these vast concerns have earned, if they have not received, the gratitude of the Government and the country for their magnificent services.

It is most unfortunate that there is no authoritative record of their work. Their achievements have been recorded piecemeal, but no attempt has been made to present them in a complete narrative, as has been done, for instance, in the case of the Mercantile Marine.

Between the years 1917 and 1919, some 4,000 vessels were dry-docked and repaired, and about 6,000 vessels repaired outside the dry docks, of all sizes and tonnage. For the greater portion of this period, the German submarine campaign was at its worst. There were from 50 to 60 vessels on stem, and from the time that the Admiralty issued instructions for dazzle-painting, arrangements were put in force in the various dry-docks for painting two or three dozen vessels per week.

Therefore, nothing could be more calculated to enhance the prestige of the channel ports, or to attract public attention to the immense possibilities of the local ship-repairing and engineering industries than a complete statement of their War work.

Even life-long residents in South Wales have no conception of its magnitude, and up to the present; the Government has failed to appreciate it at its full value.

Apart from personal matters, it is important that the work of the repairers and engineers should be more widely known, and it is inconceivable what could be more effective than the plain straightforward record of how thousands of vessels were kept at sea, for the purpose of food and munitions during the War, by the untiring efforts of the ship repairers of the Bristol Channel.

PATENT FUEL INDUSTRY.

By GUY de G. WARREN, M.Inst.M.E., F.G.S.

INTRODUCTION.

THE term "Patent Fuel" is used generally to indicate artificially moulded fuel resulting from the compression of small coal mixed with a binding substance. In this district the term is more especially applicable to the large rectangular blocks produced, as distinct from the egg-shaped forms which are known as eggettes or ovoids.

Practically the entire output of the United Kingdom is produced in South Wales, and of this 90 per cent. is exported.

The industry has become an important one in this district and the rate of progress has been constant if not rapid, having developed from the old practice of hand moulding balls of anthracite dust united by clay, which constituted the first form of patent fuel. This old method is still carried on in the western parts of South Wales, where the mixture (locally known as "pele") represents a staple form of fuel.

The production of patent fuel on a manufacturing scale was begun about sixty years ago, but even as far back as 1789 an English patent was obtained for the preparation of moulded cakes of coal mixed with clay, cowdung, broken glass, sulphur, oilcake, and pitch.

The first factory was erected in France in 1842, which was followed by the building of a factory in England in 1846.

PRODUCTION.

The progress from this date appears to have been continuous, and the industry in France, Belgium, and England had assumed large proportions when, in or about 1885, Germany began the production of patent fuel or coal briquettes, and progress in that country was so rapid that in 1913 the output was nearly equivalent to the total of all other countries combined.

The production in the United Kingdom has increased from approximately one million tons in 1885 up to about 2¼ million tons in 1913 (the last normal trading year), whereas in Germany the output of 140,000 tons in 1885 increased to nearly eight million tons in 1913.

It is also interesting to note that in proportion to the total production of coal the output of patent fuel is far less in this country than in the following European countries, where it is found that the ratios of fuel and coal outputs are :—

In Belgium—	1 ton Patent Fuel to		9 tons of coal.	
In France—	1	,,	,, 15	..
In the United Kingdom—	1	,,	148	,,

Whereas in the United Kingdom the ratio is one ton of Patent Fuel to 148 tons of coal output.

Calorific Value.

Patent Fuel is adaptable to nearly all uses to which large coal is now put. Its calorific value is somewhat higher than that of the coal used in its construction, and it has the additional advantages of compactness and ability to withstand weathering and rough handling.

Shape.

The shape adopted generally in this district is a rectangular block varying only in dimensions. The fact that practically the whole output is exported influences the design of the block, as this form lends itself to easy and economical stowage and storage, and the rubbing surfaces exposed are small in proportion to the cubical contents. The blocks manufactured vary from 9 to 26 lbs. in weight, and no definite standard of weight appears to be required by consumers, although modern practice inclines if anything towards the production of larger sizes.

Blending.

The variety of coals mined in South Wales enables fuel manufacturers to vary the composition of their product to suit requirements. With coals varying from the bituminous seams of Monmouthshire, containing up to 35 per cent., to the anthracite seams of Carmarthenshire, having 5 per cent. of volatile matter, the blending can be adapted to a variety of conditions.

The increasing demand for Patent Fuel will undoubtedly result in more attention being given by manufacturers to quality, and the necessity for eliminating ash is tending towards the introduction of coal washing plants at the works, by means of which there will be a reduced ash content of the fuel, and therefore increased value by improved quality, and in addition it means a reduced proportion of the necessary binder.

Constituents.

The materials used in the manufacture of Patent Fuel consist approximately of 90 per cent. coal and 10 per cent. coal tar pitch. Although a variety of binding substances have been tried, coal tar pitch is recognised as being the best up to the present, but efforts are constantly being made to find a cheaper material.

Process of Manufacture.

The process of manufacture varies only in minor details, and shortly stated it is as follows :—

(*a*) In works where washeries are installed, the coals are delivered from railway trucks into receiving bins and are elevated to the washeries, and after being freed from impurities are graded into nuts, beans, and fines. The former are loaded direct into railway wagons for sale, whilst the fines are passed to driers for reducing the moisture content to about 3 per cent., and are then conducted into bins.

(*b*) Pitch is delivered into a cracker, and then to a disintegrator, and afterwards is automatically delivered on to a conveyor in measured quantities, where it is joined by the fines drawn from bins (see (*a*)).

(*c*) The mixture is then passed into disintegrators for reducing the sizes and thoroughly mixing the coal and pitch.

(*d*) Elevators then convey the prepared material to pugheaters heated by superheated steam, where the fusion of the pitch is effected and the mixture becomes agglomerated in a plastic state.

(*e*) The agglomeration is then partially cooled, afterwards enters the moulds and is finally mechanically pressed and then delivered by bands for cooling and hardening on to trolleys or storage dumps.

The following diagram illustrates this:—

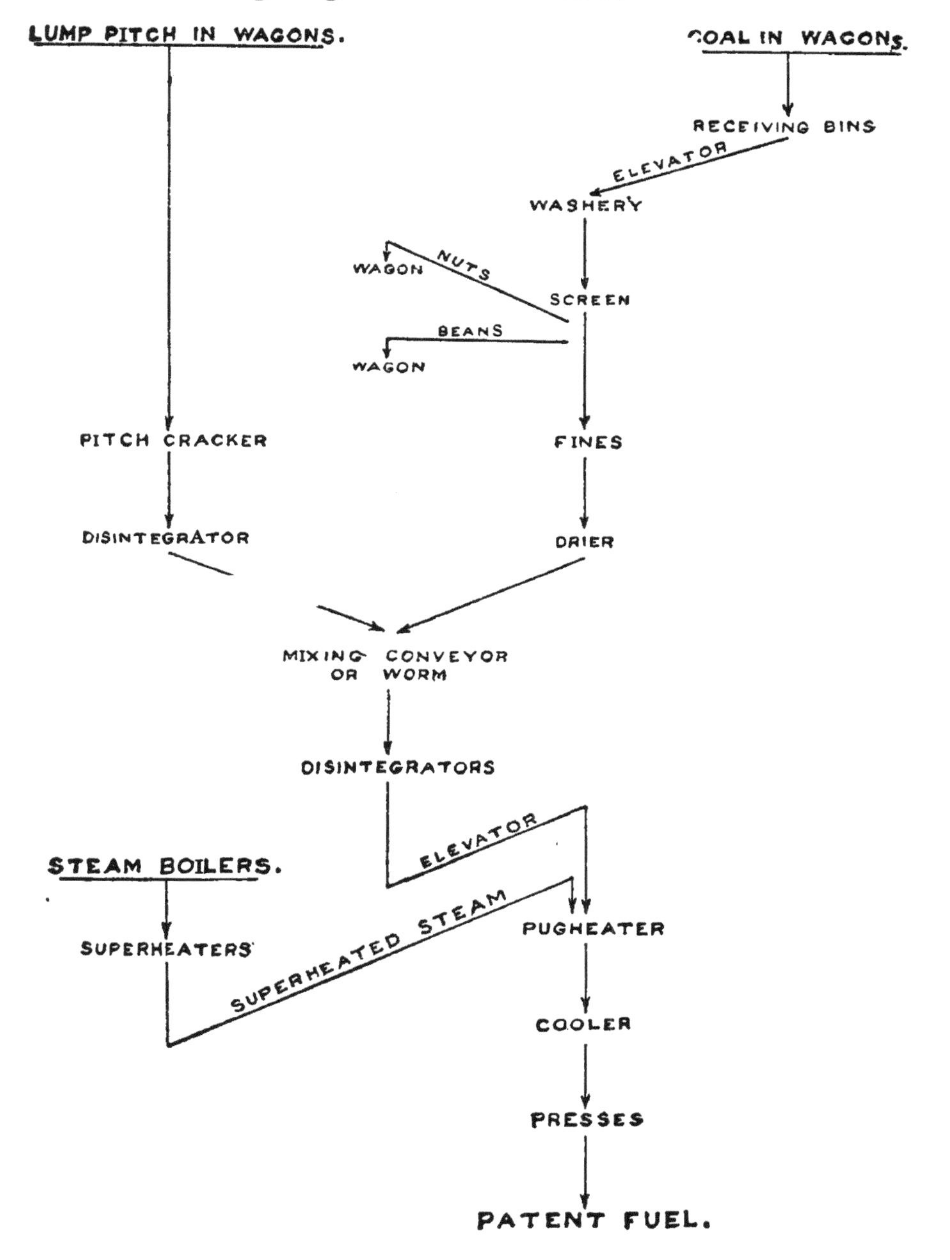

Sites of Works.

The sites of the works in South Wales are invariably chosen at or near the shipping ports, and in many cases they are erected on the dock side, so that the fuel can be directly loaded on to steamships without any intermediate railway transport.

Consumers.

The chief consumers are South American, French, Italian, and Spanish railway and steamship companies, and now that British railway companies seem to be realising the value and economy of Patent Fuel, in due course the home market will probably require increasing quantities.

Eggettes.

A limited quantity of Patent Fuel (known as eggettes or ovoids) formed in egg roll presses is made in the district. The composition and process of manufacture of the fuel is very similar to that of the rectangular blocks, but the method of pressing is effected by means of horizontal rollers rotating in opposite directions. Half-moulds are cut on the surface of the rolls, and the plastic mixture for compression is fed over the middle of the rolls, becoming shaped by the slow rotation of the rolls.

This form of fuel is well adapted for domestic use, and the demand for it appears to be increasing. This method is not more generally used, as the cost of outlay and production compares unfavourably with that of the larger rectangular blocks.

Briquetting of Anthracite.

Whereas the majority of coals worked in South Wales can be successfully briquetted, the fines of the anthracite seams can only be utilised to a very limited extent, for the simple reason that in briquettes formed only of anthracite and pitch, except when used in slow combustion stoves, the pitch burns away before the anthracite ignites, leaving a disintegrated mass of anthracite dust, which is useless as a fuel.

It is a regrettable fact, in view of the large stocks of anthracite dust, that a successful method of briquetting it has not been invented, but until a binding substance is found which will not volatilise before the coal or until some new system of coking a mixture of anthracite and pitch under pressure can be introduced, the inclusion of anthracite in the Patent Fuel of the district will be limited to its present small proportions.

Modern research has been directed towards the re-introduction of this latter process, which was originated by Warlich in a factory at Swansea about 50 years ago. There is reason to hope that this method will be rendered commercially possible within the near future.

OUTPUT AND EXPORT.

The total production of Patent Fuel in the United Kingdom in 1913, amounted to 2,213,305 tons, and of this, 2,118,910 tons were manufactured in this district.

87,762 tons of this latter quantity were consumed in this country, and 2,031,148 tons were exported, divided as follows :—

From Swansea	924,801 tons.
,, Cardiff	716,899 ,,
,, Port Talbot	242,381 ,,
,, Newport	147,067 ,,

FUTURE DEVELOPMENT.

The great reduction in the output of coal necessitates action in the direction of coal economies. This means that a more profitable use of small coal, which forms a large and increasing proportion of the total production must be made, and unquestionably one of the most economical uses is by the manufacture of Patent Fuel, representing the conversion of small coal into large coal.

Further development of the industry will inevitably take place, accompanied by improvements in the quality of the fuels produced and by the opening up of fresh markets and of additional uses to which the product of this important industry can be put.

PORTLAND CEMENT INDUSTRY.

By F. F. MISKIN, A.I.C., F.G.S.

THE greater portion of the coast of Glamorgan consists of almost vertical cliffs—in parts over 100 feet high—of Blue Lias Limestone and Shale. These rocks occur in alternate bands varying from one inch to two feet in thickness. The Limestone is well jointed and the Shale weathers readily: consequently by the action of the elèments, blocks of Limestone are perpetually falling from the cliffs, and the beach is covered with large round limestone pebbles that have been rolled by the waves of the Channel. At the mouth of the River Ddaw there is a great accumulation of Lias Limestone pebbles. These were found by Smeaton (the Eddystone Lighthouse Engineer) in 1756 to have hydraulic properties when burned to lime. The hydraulic properties are due to the fact that the Lias Limestone contains from 8 to 18 per cent. of silica and from 2 to 4 per cent. of alumina, and when burned, calcium aluminates and silicates are formed. which cause the lime to set to a hard mass after mixing with water and sand. Portland Cement is a hydraulic material of much superior quality and, unlike hydraulic lime, is non-expansive when mixed with water. Seven days after being gauged, it attains a tensile strength of 800 lbs. per square inch. Although hydraulic mortars were made in 1756, it is only since 1888, when the Penarth Cement Works were established, that the scientific development has been carried out in South Wales.

Portland Cement in this district is made from a very intimate mixture of finely ground Blue Lias Limestone and the chemically equivalent amount of aluminium silicate (Lias Shale) to satisfy the lime of the limestone. The Penarth Company was one of the pioneer firms to use the rotary kiln for burning the raw material, two of which were installed, each being 25 feet long and 5 feet diameter. This process was superseded by that of the intermittent (chamber) kilns, which in turn were supplemented by continuous vertical shaft (Schneider) kilns, and again these have been supplemented by a rotary kiln 230 feet long and 9 to 10 feet diameter.

For the fixed kilns (chamber and Schneider) the accurately proportioned prepared raw material is moulded into briquettes and burnt until incipient vitrification takes place, the resulting mass, technically known as clinker, being then finely ground, when it becomes Portland Cement of commerce.

For the rotary kilns the proportioned limestone and shale are broken down in large crushers into pieces the size of a walnut or less, then ground with water in large mills called "kominors" (*i.e.*, steel drums, each charged with 5 tons of cannon balls) from which it issues in the form of a gritty liquid. The gritty liquid is further ground in a tube mill (a long steel cylinder lined with quartz and charged with flint balls) which gives it a creamy consistency. It is then finished in a short steel cylinder, containing steel pellets as grinding bodies. 97 per cent. of this "slurry" will pass through a sieve having 32,400 meshes to the square inch. As it contains 36 per cent. of water, the slurry is stored in large masonry basins, where it is continuously automatically stirred to prevent settling. Powerful plunger pumps raise the slurry to the inlet end of the rotary kiln, which is a steel tube lined with special firebricks.

The kiln is mounted at a slight angle and rotated slowly on its own axis. The burning is effected in the kiln by the combustion of finely ground coal, blown in by an air fan at the lower end. A temperature of from 2,500° to 3,000° F. is obtained. The heated products of combustion pass up the kiln and meet the liquid slurry flowing down, causing it to dry and break into small nodules, which continue their way into the "burning zone" where they are transformed into dense hard semi-vitrified pellets called clinker. The clinker escapes from the kiln white hot, and is shot into a large revolving tube called a "cooler" through which cold air is forced. In cooling the clinker the air becomes very hot and is used in the kiln to burn the coal and save fuel.

The clinker, after storage in large silos, is conveyed to the cement mills, where it is coarsely ground in kominors or ball mills, and reground in tube mills so fine that 95 per cent. will pass through a sieve of 32,400 meshes to the square inch.

The foregoing describes the operations of the South Wales Portland Cement and Lime Co., Ltd., at their works at Lower Penarth. It is also interesting to note that the Aberthaw and Rhoose Portland Cement and Lime Co., Ltd., and the Aberthaw and Bristol Channel Portland Cement Co., Ltd., have factories situated at Rhoose and Aberthaw respectively, operating on the rotary principle, and the description of the rotary as set forth as being in operation at the South Wales Portland Cement and Lime Co., Ltd.'s works, is similar in the case of the other companies named.

In 1895 the output of cement in South Wales was approximately 5,500 tons. To-day the quantity exceeds 200,000 tons. The whole of the cement is made in close contiguity to the shipping ports of Cardiff, Penarth, and Barry, and there are excellent facilities for transport either by land or sea. The industry is carried on with unflagging enterprise typical of the busy district in which it is located. The reputation of the Cement made in the district, all made to British Standard Specification, is firmly established.

SILICA AND FIREBRICK INDUSTRY.

By P. N. F. SHEPPARD, B.A.

. SOUTH Wales is singularly well supplied with refractory materials, and this is specially the case with regard to that employed in the manufacture of Silica brick. The Millstone Grit of the Carboniferous system, which is used by all makers, crops out almost continuously round the edge of the South Wales coal-basin, between the Carboniferous Limestone and the Coal Measures ; from Monmouthshire to Pembrokeshire the visible outcrop has a length of little less than 140 miles.

Early in the nineteenth century the Silica Brick was invented by Mr. W. Weston Young, who obtained the material from quarries near the Dinas Rock at Pontneathvaughan, which quarries are still in operation ; and the name "Dinas" is well known the whole world over as denoting a silica brick of the highest quality. It is worthy of note that the Germans were not backward in adopting the name for their silica bricks.

There are thirteen works in all engaged in the manufacture of Dinas bricks, which are used in steel, copper, glass, gas, and various smelting works in this country and in all parts of the world.

Large extensions to nearly all the works in the district were carried out during the war, at the urgent request of the Ministry of Munitions of War, who were faced with a serious shortage of silica bricks, and a consequent shortage of steel and other war materials of vital importance.

The districts in which silica stone and sand are worked are as follows : Dowlais, Penderyn, the Vale of Neath, Penwyllt (Craig-y-Nos), the Black Mountain, Llandebie, Kidwelly, and Templeton, near Narberth.

The stone, which in former days was as a rule procured from underground workings, is now almost entirely quarried in the open. According to the measurements of the Geological Survey, the quarry face at the Bwamaen and Kilhepste workings, at Pontneathvaughan, near Glyn Neath, is nearly 100 feet high, and the seams or beds of silica are from 16 to 20 feet in thickness.

The following are two typical analyses of Silica rock :—

		(1)	(2)
(1)	Silica	98·31	96·73
	Alumina	·72	1·39
	Protoxide Iron.. ..	·18	·48
	Lime	·22	·19
	Potash and Soda ..	·14	·20
	Water	·35	·50
		99·92	99·49

The methods employed in the manufacture of silica brick are much alike in all the works. After having passed through one or more crushers, the stone is fed into a grinding mill, which is, as a rule, a circular pan with a solid bottom and revolving runners edged with steel or iron.

Lime for binding the silica rock is added during the grinding, usually in the form of milk of lime. When the material has reached a suitable degree of fineness, it is removed from the mill and placed on the tempering floor, where it is kneaded under foot preparatory to moulding.

The bricks are made by hand or by hand-press or power press, and then placed on a heated drying floor or in a tunnel drier. After drying for a night, the bricks are fired in updraught beehive kilns for about a week, a very high temperature being obtained, and the bricks are allowed to cool very slowly before being drawn out of the kiln.

Silica sand is worked in some districts, and used for furnace bottoms and for setting the bricks ; and at several works the rock is ground to various degrees of fineness commonly known as "ganister" and "silica cement" when in a dry form, or as "ganister mixture" and "ground ganister" when in a plastic state.

In some cases also the black shales which occur between the beds of silica are utilised, and from the analysis of some of these deposits there would seem to be further scope for the development of this branch of the industry.

With regard to the firebrick trade, there are considerable deposits of fire clay of varying quality throughout the South Wales Coalfield.

In many places the clay is worked in the ordinary course of coal getting and tipped out as rubbish, whereas it is in fact a valuable commodity. The majority of the firebrick works are of comparatively small size, and make a brick of good ordinary quality for local consumption. Messrs. Guest, Keen & Nettlefolds manufacture a good class firebrick, which, for the most part, is utilised in their own steel and other works.

There are, however, several firms who make a brick approximating to the Stourbridge quality, amongst whom may be mentioned the Blaendare Company, Pontypool, the Darren Brickworks, and Messrs. Southwood Jones & Co., Ltd., of Risca.

The latter firm has a capacity of about 5,000 tons per annum, and besides executing work for the British Admiralty, has regularly exported to many distant markets, including Mexico, Russia, Australia, and India.

Messrs. Southwood Jones & Co. also work Magnesian Limestone, or Dolomite.

It will be seen, from these short particulars, that the South Wales user of refractory materials has little need to go outside his own district to supply his requirements.

AGRICULTURE.

By HUBERT ALEXANDER.

AGRICULTURE in the County of Glamorgan must of necessity be considered of far less importance than many other industries, particularly the coal industry, which has made South Wales what it is to-day. It is doubtful, however, whether, in so small an area, the difficulties of cultivation are so pronounced, or vary to such an extent, as in this county, for the reason that within a comparatively short distance are found the fertile valley known as the Vale of Glamorgan, extending from the base of the hills to the Bristol Channel, and immediately at its back the hill districts, lying at an elevation of from 700 feet to 1,200 feet above sea level, thus rendering two different modes of farming necessary in the respective areas. While the growing of cereal crops is practised extensively in the Vale, the hill farms are devoted to the rearing of stock, particularly Welsh mountain sheep, which predominate in this area. The soil varies, in some places being principally of Mountain Limestone, and in others Old Red Sandstone, the former formation running along the seaboard from Cardiff to Porthcawl, where a number of large Cement and Lime Works are at present in vigorous operation.

One of the features of farming in the county is the predominating number of what may be termed Small Holdings, it being the exception to find any holding in excess of 150 acres in extent. There are also a large number of occupying owners. The farmhouses are, as a rule, poor as compared with the farmhouses in many of the English counties, but the mode of farming by the larger farmers is up-to-date in every respect, all modern improvements in machinery and labour-saving devices being employed to the fullest extent. In this connection it should be pointed out that the difficulties of labour, on account of the great competition of mines, docks, and other important industries in the county, render the labour question a paramount one, and on account of the large wages earned in the industrial centres, the problem of retaining the agricultural labourer on the land is one of the greatest difficulty and importance. There is a great shortage of cottages for agricultural workers, and the cottages in many instances are such that it is almost unfair to ask a labourer and his wife to live and bring up a family where modern comforts and conveniences do not exist. The older men are content to continue with their old employers under similar conditions to those under which they have been in the habit of living, but the younger men, especially upon their return from Active Service, are disinclined to continue to live under such conditions, and the housing problem therefore is one of the most important causes of rural depopulation in this county.

As might be anticipated, where there is a very large industrial population, dairy farming is practised to a considerable extent in the county, and a large quantity of milk is produced in the immediate neighbourhood of all the large cities and towns, but the supply is far from being sufficient for the requirements, with the result that quantities of milk are daily received by rail from Carmarthenshire and West Wales, and also from Somersetshire and Gloucestershire.

Stock rearing is perhaps the most important section of agriculture practised in the county, and this is accounted for by the fact that there is an unlimited market at the farmer's door for stock of every description. A number of the larger farmers successfully breed pedigree stock of the highest quality, there being a number of prominent pedigree herds of Hereford and Shorthorn cattle, Shropshire sheep, and shire and hackney horses, and Welsh ponies in the county. The latter are bred almost in a wild state on the commons in the Peninsula of Gower.

The production of corn, in view of the small available area suitable for corn growing, is above the average, and it is interesting to note the effort made by agriculturalists in the county, during the War, in increased corn production, as compared with the remainder of England and Wales. The following Table, issued by the Glamorgan War Executive Committee, gives a summary of these results :—

ENGLAND AND WALES.			GLAMORGAN.	
Crop.	Acreage Increase.	Percentage of increase.	Acreage Increase.	Percentage of increase.
Wheat	753,000 acres	39½%	8,375 acres	173%
Barley	158,000 ,,	11¾%	1,709 ,,	35%
Oats	736,000 ,,	70%	13,419 ,,	121%
Other Crops (Rye, Beans, etc.) ..	280,000 ,,	46¼%	1,439 ,,	663%
Potatoes	198,000 ,,	33⅓%	1,056 ,,	45%

The Crops grown in the county include—Wheat, Oats, Barley, Mangel, Swedes, Turnips, Potatoes, in addition to which, in the vicinity of the cities and towns, there is a considerable production of vegetables from extensive market gardens.

The rainfall in the county is one that varies to a very great extent in different localities ; whereas, in the Vale of Glamorgan, from returns published in 1890, the rainfall was 27 inches during the year, in the hill districts it amounted to from 70 inches to 78 inches.

The county has a number of small Agricultural Societies, established for the encouragement and improvement of the breed of all classes of stock. It has also a vigorous Chamber of Agriculture, and during recent years a flourishing branch of the National Farmers' Union, these bodies making it their business to focus public opinion, particularly upon all agricultural legislation, with a very considerable benefit to the agricultural community of the district.

A matter of very considerable importance to the agricultural community in the county concerns the steps that have recently been taken for the erection and equipment of an Agricultural Department at the University College at Cardiff. Through the generosity of Lord Glanely, who has given a sum of £25,000, it is proposed to erect a building for the Agricultural Department, and it is further proposed that a Chair of Agriculture shall be established by a joint contribution through the trustees of a fund of £5,000, representing surplus contributions received by them in connection with the visits to Cardiff of the Royal Agricultural Society and the Bath and West of England Agricultural Society, and a similar contribution from the Treasury, which has undertaken to give a pound for every pound found locally. It will therefore be seen that the Chair of Agriculture will be endowed with a sum of £500 a year. In addition, the salaries of the Chemistry and Botany teachers and the staff expenses will be derived from a grant from the Development Commissioners, and an application is at present being made for a grant from this source, amounting to £1,000 a year. Added to this, a grant of £2,500 a year is expected from the University Court, making a total endowment of £4,000 a year.

It should be further mentioned, that in addition to this scheme, the County Council has determined to establish a Farm Institute. The whole of the elementary and practical work at present carried on at the University College, as short courses, will be transferred to the Institute, and the instruction in this direction will be consequently rearranged. Only the higher Agricultural instruction and the purely technical and scientific work will continue to be dealt with at the University College.

The University will arrange Certificate and Diploma courses, and also courses for the new degree in Agriculture formulated in Lord Haldane's Report, open to students from any county, so that in the future it will be possible to grant Certificates and Diplomas in Agriculture and Dairying to students attending the University College at Cardiff, as at other Welsh and English Colleges.

SCIENTIFIC SOCIETIES.

By GILBERT D. SHEPHERD, F.C.A.

THE first general Scientific Society to be established in Cardiff was the "Glamorganshire and Cardiff Literary and Scientific Institution," which was founded on 30th October, 1837, under the presidency of the Marquess of Bute, "to form a collection of the best books on science and literature, and a museum for the deposit of such objects in Natural History, Science and Art as may be presented to the Society." The annual subscription was one guinea, and the first year's income amounted to £162. The Institution occupied two rooms at the Cardiff Free School, but afterwards moved to No. 3, Crockherbtown. The Marquess of Bute, The Very Rev. Dean Conybeare, F.R.S., The Rev. J. M. Treharne, and Captain W. H. Smyth, R.N., were the principal donors to the Museum.

After a somewhat chequered career covering a period of 21 years, the Institution ceased to exist soon after the Annual Meeting held on 17th February, 1859. Fortunately the Museum collection was not dispersed, and a few years later was transferred to the care of the Cardiff Free Library and Museum Committee.

On 14th September, 1841, a Mechanics' Institute was formed with the main object of arranging Lectures on scientific subjects. The first two lectures were on "Druidical Remains" and "The Human Eye" respectively. The Institute was reconstituted on 18th March, 1848, under the title of "The Cardiff Athenaeum and Mechanics' Institute." In 1851 it removed to Church Street, but its later history is not known, although it is believed to have been closed down about 1856.

In 1858, preliminary steps were taken to establish a Free Library and Museum for Cardiff, but at a public meeting held in 1860 a proposal to adopt the "Free Libraries

Act, 1855," was defeated. A subscription list was at once started for a voluntary free library, and a reading room and news room was opened on 14th June, 1861. The old museum collection of the Literary and Scientific Institution was transferred to the Library Committee. No use was made of it, however, until 1865, when bigger premises were taken, and £43 was spent on cases in which the museum could be displayed in a separate room appropriated for that purpose. In the meantime (1862-3) the Libraries Act had been adopted, and the Free Library Committee was working under the Cardiff Corporation, with rate aid. On 8th December, 1863, a Sub-Committee numbering eight was appointed "to manage and arrange the Museum," but there is no record of their work later than 30th June, 1865.

In 1867 was founded the Cardiff Naturalists' Society, which celebrated its Jubilee three years ago and is still flourishing. The objects of the Society as originally constituted were "the practical study of Natural History, Geology, and the Physical Sciences, and the formation of a Museum in connection with the Free Library," but these were extended on two occasions, and the Society has now become much broader than its name would appear to indicate. Starting with a membership of 24, the Society has made more or less steady progress, until it now numbers 585 members, with 650 holders of Family Tickets. Annual Transactions have been issued since 1867, and much original work has appeared therein. The Society has published "The Birds of Glamorgan" and "The Flora of Glamorgan" (two editions), and is now compiling "The Fauna of Glamorgan" much of the material for which has already appeared in the Transactions. It has conducted excavations at the Roman Fort of Gellygaer and the Roman Villas at Llantwit Major and Ely, and has supervised the excavation of "Castell Morgraig." Other antiquarian researches cover the Gower and Doward Caves, the St. Nicholas Cromlechs, the pre-Norman Inscribed Monumental Stones of the County, the Calvary Crosses of Glamorgan, Old Maps, Old Wills, the Cardiff Charters, etc.

Lectures are given fortnightly during the winter months, and Field Meetings are held in the summer. Useful work is undertaken by the Sections—(*a*) Biological and Geological, (*b*) Archaeological, and (*c*) Photographic. As one of the original Corresponding Societies of the British Association, it has usually been represented at the Annual Conferences of the Corresponding Societies.

The Society has collated the Meteorological Returns each year since 1867, and has presented several recording instruments to the Cardiff Meteorological Station, including Seismograph, Sunshine Recorder, Solar Radiation Thermometer, etc. It exchanges its publications with a large number of other Scientific Societies and possesses a large Library, which is at present located in the Cardiff Central Library.

From its foundation, the Society has been closely associated with the Museum, and it assisted the old Free Library Committee in all possible ways. From 1876 to 1912, members of the Society were co-opted on the Museum Sub-Committee, and it is acknowledged that much of the success of the Cardiff Museum was due to the active co-operation of the Society.

It may truly be said that the nucleus of the present National Museum of Wales was started in 1837 by the establishment of the Museum of the Glamorganshire and Cardiff Literary and Scientific Institution. As already mentioned, this was transferred soon after 1859 to the Free Library and Museum Committee, and in turn the Museum built up by the latter was handed over to the National Museum of Wales in 1912.

Agriculture has always held a foremost place in the County, and the first Society of which any record can be traced is the Glamorgan Agricultural Society, which was established at Cowbridge on 28th October, 1772. During its career it had a considerable influence in increasing production by improved draining, better crops, and a higher standard of stock. The Society unfortunately lapsed in 1903.

On 8th January, 1842, the Cardiff Farmers Club was formed "for the purpose of disseminating sound practical information amongst the cultivators of the soil," but this did not live for many years.

The Glamorgan Chamber of Agriculture was established on 12th January, 1904, and has served a very useful purpose in the encouragement of scientific agriculture.

On 22nd September, 1828, was founded the Glamorgan and Monmouthshire Horticultural Society, whose history it has not been possible to trace. In 1862 a similar Society was started under the title of the "Glamorganshire Horticultural Society," and this was in existence for a number of years.

The South Wales Institute of Engineers was founded in 1857, and incorporated in 1881. It has succeeded in placing Engineering and Coal Mining in this District on a high level, as shown by the Transactions of the Institute.

The Cardiff Photographic Society was established in 1886, and had a useful career for 24 years. In 1890 it presented to the Cardiff Central Library 1,200 photographs and a number of lantern slides of Castles, Churches, Crosses, etc., in the County, and this may be regarded as the first step towards a Photographic Survey of the District. The Society, through death and otherwise, gradually dwindled in membership and became defunct in 1910. In the following year, steps were taken to form a Photographic Section of the Cardiff Naturalists' Society, and this was established on 25th October, 1911, immediately securing a large membership, which it has maintained. Through the instrumentality of the Section, a systematic Photographic Survey of Wales and Monmouthshire has been developed in connection with the National Museum of Wales, but unfortunately the war has interfered with its progress. It is hoped that this may shortly be taken up seriously by amateur photographers in the district, as there is ample scope for useful work in this direction. In 1912, a Federation of the Photographic Societies of Wales and Monmouthshire was formed, and this includes Societies at Cardiff, Newport, Barry, etc.

The Astronomical Society of Wales was founded in 1895 with a membership of 60, and during the intervening 25 years has done much useful work, including the publication of its Journal and the collection of Astronomical books and drawings. The latter have been deposited in the Cardiff Central Library and are available to the public.

In addition to the foregoing, there are Societies attached to the University College (Geographical, Engineering) and to the Technical College (Engineering, Chemical and Economics).

There are also various professional Societies (dealing with Medicine, Economics, Education, etc.) much of whose work is on scientific lines.

METEOROLOGY OF CARDIFF AND DISTRICT.

By EDWARD WALFORD, M.D.

THE meteorological conditions of Cardiff correspond closely with those of South Wales generally, and do not present any very striking or special features of interest. The climate is mild, neither very hot in summer, nor very cold in winter, and may be described as temperate, both extremes being subject to the moderating influence of the sea and estuary of the Severn in the Bristol Channel. This Channel measures about 8½ miles in width at Lavernock Point, widening out to 75 miles opposite St. Gowans. Some of the currents of the Gulf Stream enter the Bristol Channel and exert a considerable influence upon the climate of the whole of the coast of South Wales, contributing in no small degree to the comparative mildness and humidity which prevail in this area. The mean annual temperature of the coast at St. Davids is 49·1° F., at Aberystwyth 51° F., and at Carmarthen and Cardiff, being more inland, 48·8°. In the first quarter of the year, the mean is approximately two degrees warmer than on the East Coast of England. The mean of the second quarter differs little from that of the East Coast. In the third quarter, the mean of the East Coast is some degrees higher than that of South Wales, whereas the mean of the fourth quarter is considerably above that of the East Coast. These records point, therefore, to the influence of the Gulf Stream and sea on the Welsh Coast, in lowering the summer heat and in checking the fall of temperature in the winter.

The rainfall of Cardiff is of course greater than on the East Coast, but less than at some places further west, or in the proximity of the hills to the north of the town. On the whole, a plentiful rainfall is an advantage in a large city as a useful auxiliary to the cleansing operations of the municipality, and a means of removing organic dust and micro-organisms from the atmosphere. The prolonged deficiency of rain during the summer of 1919 was one of the marked features in the meteorology of the year, and resulted in some temporary inconvenience, by limiting the supply of the excellent water from the reservoirs in Breconshire at a time when water for municipal and other purposes was much required. Owing to the above-mentioned conditions, the relative humidity, estimated from the readings of the hygrometer (wet and dry bulb thermometer) is high. In this connection the researches of Dr. Leonard Hill are of great interest. In a lecture which he delivered before the Meteorological Society in March, 1919, he points out that "it is the cooling and evaporating power of the atmosphere and the radiant heat of the sun or other sources of radiant energy, which have a colossal effect on our comfort and well-being, and it is these factors which require to be measured by the student of hygiene." He is of opinion that the cooling power of the air, under specified conditions depending upon wind and evaporation, is a very important property of the atmosphere from the hygienic point of view, and that this can be determined by an instrument which he calls a Kata-thermometer, *i.e.*, an ordinary thermometer, wet or dry, raised to the temperature of 100° F. and exposed, the time taken for a fall of 5° being accurately read in seconds with a stop-watch, and the average of several readings taken. Dr. Hill considers that the use of this instrument is of more value from a hygienic point of view than the records of relative humidity deduced from the readings of the wet and dry bulb thermometer, as these do not give the factors which determine comfort and health. It is interesting to note that this observer remarks that the trouble and expense of artificially moistening air after warming it in plenum systems of ventilation is wholly wasted, and that it is the low cooling power, the monotonous touch of warm air and absence of radiant energy which condemn this system and which give the opposite of the ideal conditions, which are warm floor, cool variable airs, and a genial, radiant heat. With these observations most of those who have had experience of the plenum system will cordially agree. The daily and monthly readings of the meteorological instruments in this district are not usually characterised by any remarkable variety. The normal and uneventful records are occasionally relieved by the occurrence of an earthquake recorded on the Seismograph (an instrument which forms part of the equipment of the Station), but as these disturbances generally occur at a distance of some thousands of miles, no undue excitement is caused thereby.

The geographical position of the City of Cardiff, and the geological formations, especially the superficial strata of the neighbourhood, exert of course a powerful influence upon the climate and meteorology of the district. The town lies at the mouth of three rivers, the Rhymney on the eastern boundary, the Ely on the west, and the Taff passing through the centre of the town, all three discharging their contents into the Bristol Channel. The low-lying alluvial plain, upon which part of the town is built, stretches from the western boundary along the course of the Severn to Chepstow. The lower or southern parts rest upon a stiff blue marine clay of considerable thickness near the Docks, which becomes thinner and gradually terminates about the centre of the town, the underlying gravel coming to the surface and extending in a northerly direction with varying thickness, attaining a height of about 40 feet above mean sea level in Queen Street and Newport Road, to about 90 feet in Cathays, and resting upon the New Red Marl, the Old Red Sandstone and Silurian formations cropping up on the Penylan Hill at an elevation of about 200 feet, a range of hills about 5 miles to the north of Cardiff forming the southern edge of the South Wales Coalfield. Dr. F. J. North,* of the Geological Department of the National Museum of Wales, states "that these gravels belong to the post-glacial terrace gravels deposited by the Rivers Taff and Rhymney. Such gravels underlie

* *Transactions of the Cardiff Naturalists' Society*, 1915.

the greater part of Cardiff, east of the River Taff, and resting as they do upon the impervious Keuper Marl, hold much water. They were at one time an important source of water for the City, but the water is subject to pollution and is not now used for domestic purposes." Doubtless the great improvement of the public health of the district is largely due to the substitution of a pure water supply from the Brecon Hills in the place of this gravel water. The warming and humidifying influence of the sea and Gulf Stream and the moist South-west winds is naturally more marked on the coasts of Pembrokeshire and Carmarthenshire, and diminishes towards the east, being hardly felt at all in the Gloucestershire part of the Bristol Channel.

The Cardiff Municipal Meteorological Station, at which the following observations were taken, is situated on Penylan Hill at a height of 203 feet above Mean Sea level, Latitude 51°30 N. and Longitude 3°10 W. It is recognised by the Meteorological Office as a normal Climatological Station at which the readings are taken at 9.0 a.m. and 9.0 p.m. It is equipped with the usual instruments. The dry and wet bulb thermometer (Hygrometer), a Phillips' maximum thermometer, and a Rutherford's spirit minimum thermometer are placed in a double-louvred Stevenson's screen of the pattern approved by the Royal Meteorological Society, four feet above the ground, with the door opening to the north. On the land adjoining the screen, and fenced in with railings, are other instruments. Two thermometers for taking earth temperatures placed at one foot and four feet respectively below the surface. A Solar radiation maximum thermometer mounted on a post four feet above the ground ; this is a black bulb thermometer in vacuo, and is freely exposed to the Sun, with the bulb directed to the South. The terrestrial minimum thermometer is fixed horizontally on short grass, and is contained in a glass shield hermetically sealed. The rain gauge is of the Snowden pattern, the diameter of the cylinder being five inches, the rim of the gauge being twelve inches above ground. These instruments are placed in an enclosure 25 feet square, surrounded by railings. The sunshine recorder in use is of the Campbell Stokes pattern, and is placed on a coping of the Penylan reservoir in such a position that the sun can shine on it the whole time he is above the horizon. A wind vane is fixed on the top of the water tower, and the direction of the currents is duly recorded in the returns. In a separate building, in the base of the water tower, is placed a barometer of the Kew pattern, provided with an attached thermometer and a double scale and vernier divided on one side into inches, ·10, ·05, and ·002 inches, and the other side into millimetres, etc. The instruments have all been verified by the Kew Observatory, and the necessary instrumental corrections are duly made. Weekly and monthly reports on the weather are sent to the Meteorological Office, South Kensington, which are included in the returns issued by that Office. As already mentioned, a Seismograph forms part of the equipment of the Station. This interesting and instructive instrument was presented to the Cardiff Corporation by the Naturalists' Society in 1909, and is of the Milne horizontal pattern. It stands on a cement concrete column eighteen inches square, three feet above the floor level.

The Meteorological Station is under the control of the Obesrvatory Committee of the Cardiff Corporation (under the chairmanship of Councillor Sydney Jenkins), which displays a keen and generous interest in all that concerns local meteorology and which has at the present time under consideration the question of extending the scheme of observations so as to include continuous records of atmospheric pressure, temperature, wind, sunshine, and rainfall, etc., by means of self-recording instruments and of converting the Station into one of the "First Order" of the International classification. The Medical Officer of Health is the responsible observer, and is aided in taking the readings by two assistants, one of whom is a member of his staff, the other, lent by the Waterworks Engineer, is the resident Inspector, who lives in a house attached to the Penylan reservoir.

Rainfall.

Records of rainfall in Cardiff extend over a period of 55 years, although the position of the rain gauges has been altered from time to time. The observations in the years 1908-1918, inclusive, were made at the Penylan Meteorological Station, but the slight change of position of the gauges hardly affects their accuracy. The average rainfall in Cardiff for these 55 years was 42·86 inches. This compares with 39·16 inches at the Lisvane Reservoir for the same period, and with 40·61 inches, the average for 32 years at the Heath Filters, both these gauges being within four miles of Cardiff. In the several drainage areas of the Cardiff Waterworks, the rainfall varied as follows. In the Taff Vawr Section at Nant Penig at an elevation of 2,000 feet, the average for 24 years was 84·51 inches. At the Cantref Reservoir, at an elevation of 1,120 feet, the average for 30 years was 69·40 inches. At the Llwynon Reservoir, at an elevation of 860 feet, the average for 10 years was 62·14 inches. In Cardiff in the 55 years, the maximum of 56·73 inches occurred in 1882, and the minimum of 31·59 inches in 1887. The average monthly rainfall in Cardiff was :—

January	February	March	April	May	June
in.	in.	in.	in.	in.	in.
3.84	3.10	2.87	2.50	2.48	2.62
July	August	September	October	November	December
in.	in.	in.	in.	in.	in.
3 34	4.51	3.64	4.88	4.03	4.60

The average number of rainy days was 195. It will be seen that October and December are the rainiest months, and that the rainfall in August is not much below that of December. The rainfalls in 1918 and 1919 are not included in these averages, and present certain exceptional features.

The rainfall in 1918 was :—January, 5·47 inches, February, 2·97 inches, March, 2·76 inches, April, 2·39 inches, May, 1·52 inches, June, 2·27 inches, July, 5·95 inches, August, 3·39 inches, September, 10·49 inches, October, 29·0 inches, November, 3·52 inches, December, 7·13 inches, making a total of 50·76 inches. The rainfall for 1919 is given in the following Tables.

In *British Rainfall* for 1917, the following records appear relating to Wales and Monmouthshire, and the average rainfall in several Stations in this area for the years 1875-1909. Abergavenny, 37·11 inches, Carmarthen, 48·80 inches, St. Davids, 36·75 inches, Aberystwyth, 45·46 inches, Tyrmynydd (Radnor), 62·06 inches, Tan-y-bwlch (Merioneth), 62·25 inches, Ystalyfera (Glamorgan), 64·04 inches.

If the rainfall in Wales is compared with that of London and the Eastern Counties, the effect of local conditions becomes very apparent. The average rainfall in London (Camden Square) for the same period was 25·11 inches, Tenterden (Kent), 27·64 inches, Margate, 23·33 inches, Shoeburyness (Essex), 19·28 inches. The heaviest average rainfall in the British Isles is given as 129·48 inches, in Borrowdale (Seathwaite) Cumberland.

The following tables relate to Barometric Pressure, Temperature, Relative Humidity, Rainfall, and Sunshine during 1919.

TABLE I.

Barometric Pressure and Relative Humidity.

1919	Mean Barometric Pressure *		Hygrometer *		
	Uncorrec'd.	At M.S.L. and 32° F.	Dry-bulb (mean).	Wet-bulb (mean).	Mean Relative Humidity.
	in.	in.	° F	° F	%
January	29.519	29.738	38.0	36.8	89
February	29.542	29.774	35 1	34.0	87
March	29.503	29.826	39·1	37 4	85
April	29.807	30.013	44.6	41.6	78
May	29.877	30.054	54.7	51.1	77
June	30.027	30.190	56.5	52.3	74
July	29.920	30.130	57.6	53.9	77
August	29.893	30.081	61.0	57.0	82
September	29.870	30.032	55.1	52.4	82
October	29.949	30.143	45.1	43.7	89
November	29.594	29.811	38.5	36.9	86
December	29.608	29.821	43.1	41.9	90
Means	29.759	29.967	47.3	44.9	83

* From observations at 9.0 a.m. and 9.0 p.m.

TABLE II.

TEMPERATURE.

1919	Maximum	Minimum.	Mean of Maximum.	Mean of Minimum.	Mean Temperature.	Difference from Av. (30 years)
	° F	° F	° F	° F	° F	° F
January ..	51.2	24.7	42 9	32.8	37.8	—1.6
February	49.2	22.1	39.8	31.5	35.6	—4.5
March ..	53.3	26.0	44.9	33.2	39.0	—3.2
April ..	68.0	30.0	51.8	37.6	44.7	—1.6
May ..	76.3	40.0	63.9	47.1	55.5	+2.8
June ..	78.0	43.0	65.8	49.7	57.7	+0.6
July ..	75.2	45.6	66.0	50 6	58.3	—2.4
August ..	82.5	42.0	69.2	53·6	61.4	+1.1
September	82.0	34.5	62.8	48·1	55.4	—0.9
October ..	63.0	32.5	53.8	39·4	46.6	+3.6
November	54.5	25.5	43.2	34.5	38.5	—5.9
December	52.2	31.8	47.6	38.0	42.5	+1.8
	Max 82 5	Min 22.1	Mean 54.3	Mean 41.3	Mean 47.7 ..	—1.4

TABLE III.

RAINFALL.

1919.	Amount.	Difference from av. (30 years).	*Greatest fall in 24 hours.	*Date of Greatest Fall.	* No of Days with rain (0.01 in. or more)
	in.	in.	in.		
January	5.72	+2 20	.80	14th	21
February	4.63	+1.70	1.40	5th	12
March	5.78	+2.58	.75	18th	19
April	1.93	— .71	.59	13th	15
May	.71	—1.76	.19	11th	6
June	1.32	—1.60	.64	12th	12
July	2.14	— .54	.97	19th	9
August	2.82	—1.45	.95	27th	11
September	2.20	— .69	1,02	22nd	11
October	2.64	—2.39	1.17	23rd	8
November	3.40	— .05	.77	30th	16
December	7.59	+3.01	.82	28th	26
	Total 40.88	+ .29	1.40	5th Feb.	Total 166

* 24 hours ending 9.0 a.m. next day.

TABLE IV.

SUNSHINE.

1919.	Totals, 1908 to 1918.	11 years Average.	1919.	Difference from Av. 11 years.
	hrs. & 10ths.	hrs. & 10ths.	hrs. & 10ths.	hrs. & 10ths
January	580.2	52.5	61.1	+ 8.6
February	843.3	76.6	71.8	+ 4.8
March	1193.6	108.5	105.8	— 2.7
April	2030.3	184.5	163.2	—21.3
May	2393.2	217.5	198.5	—19.0
June	2369.3	215.3	224.1	+ 8.8
July	2410.5	219.1	186.7	—22.4
August	2109.2	191.7	228.2	+36.5
September	1617.4	147.0	147.6	+ .6
October	1071.2	93.7	150.8	+57.1
November	735.2	66.8	55.9	—10.9
December	568.6	51.6	44.4	— 7.2
	17,922.0	1629.2	1638.1	+23.3

BOTANY.

By E. VACHELL, F.L.S.

The County of Glamorgan, owing to its mixed Geological character and to the diversity of its physical structure, is peculiarly interesting from a Botanical point of view.

Few counties show within similar limits such striking contrasts—moorland, bog, salt-marsh, sand-dune, and alluvial river deposit are all represented and afford suitable habitats for an unusually large number of species, while the warm damp climate produces an exceptional luxuriance of growth.

The high hill country in the north called Blaenau Morganwg, is composed almost entirely of sandstones and shales, with outcrops of mountain limestone.

The hills in this district are for the most part bleak and bare, and afford rough pasturage for sheep. They are covered with a damp peaty earth and the vegetation is, generally speaking, uninteresting, being composed largely of gorse, heather, bracken, and other typical moorland plants.

In the neighbourhood of Craig-y-llyn (1,969 feet) one of the highest points in the County, *Thalictrum collinum, Rubus saxatilis, Geum rivale, Saxifraga hypnoides, Sedum roseum, S. Telephium, S. Forsterianum, Vaccinium Vitis-Idæa, Pyrola secunda, Pinguicula vulgaris, Hymenophyllum peltatum, Cryptogramme crispa, Lastræa Thelypteris, Phegopteris Dryopteris,* and *Lycopodium Selago* are to be found, while *Lobelia Dortmanna, Littorella uniflora, Sparganium affine, Isoetis lacustris, I. echinospora* occur on the margins of the two small mountain lakes Llyn-fawr and Llyn-fach, one of which has recently been converted into a waterworks. These hills are intersected by deep valleys, which were once luxuriant with vegetation, but are now often spoilt by the smoke and dirt of innumerable collieries and furnaces. Many old records show that *Trollius Europæus* and *Hymenophyllum tunbridgense* were once comparatively common, now alas, they are almost extinct in the County.

The rocky beds of the mountain torrents that fall into the Vale of Neath form an excellent example of what many of these valleys must once have been. The damp ledges of the spray-covered rocks near the waterfalls form a natural rockwork of exceptional beauty, and are the home of *Meconopsis cambrica*, *Rubus idæus*, *R. saxatilis*, *Geum rivale*, *Chrysosplenium alternifolium*, *Crepis paludosa*, *Vaccinium Myrtillus*, *Phegopteris Dryopteris*, *P. polypodioides* and *Polystichum lobatum* var. *lonchitidoides*, a form which has given rise to many old legends of the existence of *Polystichum Lonchitis* in Glamorgan. The following new brambles are mentioned by the Rev. W. Moyle Rogers and the Rev. Augustin Ley as occurring in this district (*Journal of Botany*, 1906) *Rubus Godroni* Lec. and Lam., var. *foliolatus*, *Rubus lasioclados* Focke, var. *longus*, *Rubus ericetorum* Lefv., var. *cuneatus*, and *Rubus horridicaulis* (P. J. Muell.).

These ravines are usually clothed with trees, the most common forms being *Pyrus Aucuparia*, *Cratægus Oxyacantha*, *Fraxinus excelsior*, *Alnus glutinosa*, *Corylus Avellana*, *Quercus Robur*, *Salix caprea*, and *S. aurita*.

Two or three other rare plants from this northern district deserve to be mentioned. *Asplenium viride* still survives at Morlais Castle, *Carex dioica* occurs on the hills between Llwydcoed and Merthyr Tydfil, *Utricularia minor* grows sparingly in Penclun Marsh, and *Sibthorpia europæa* on the Garth mountain.

Between this high hill-country and the sea lies the Vale of Glamorgan or Bro Morganwg, a belt of rich agricultural land which provides extensive pasturage for cattle, an excellent soil for the cultivation of wheat, and is noticeable for the luxuriance of its vegetation.

Geologically this part of the County is composed chiefly of Lias, with tracts of Mountain Limestone. Amongst the richly cultivated land a few undrained bogs and moorlands still remain and afford excellent harbourage for wild plants.

At Mynydd-y-Glew, a wild tract of country about ten miles from Cardiff, it has been possible to sée *Radiola linoides*, *Potentilla palustris*, *Drosera rotundifolia*, *D. longifolia*, *Apium inundatum*, *Peplis portula*, *Calluna vulgaris*, *Oxycoccos quadripetala*, *Erica Tetralix*, *E. cinerea*, *Menyanthes trifoliata*, *Myosotis sylvatica*, *Veronica scutellata*, *Scutellaria minor*, *Polygonatum multiflorum*, *Sparganium minimum*, *Potamogeton obtusifolius*, *Carex leporina*, *C. echinata*, *C. curta*, *C. pilulifera*, *C. pallescens*, *C. helodes*, *C. binervis*, *C. inflata*, *Osmunda regalis*, *Equisetum sylvaticum*, and *Pilularia globulifera* all growing within a radius of half-a-mile, while *Ranunculus lingua* and *Orchis praetermissa* flourish in a similar tract of bogland at Ystradowen a few miles away.

A few plants are of special interest on account of their rarity or because of the profusion of their growth in the Vale of Glamorgan.

To the former class belong *Carex montana*, which grows extensively on the slopes of a rocky gorge near St. Brides Major, *Lathyrus palustris* and *Scirpus rufus* which have been found in Gower, *Scrophularia nodosa* var. *Bobartii*, which grows in a small pond on Sully Island, and *Liparis Loeselii* var. *ovata*, known to occur in Carmarthen Bay, which has recently been found to extend sparingly into Glamorgan. (*Journal of Botany*, 1905, p. 274.)

To the latter class belong *Clematis vitalba*, which flourishes in some parts of the Vale, covering hedges, shrubs, and trees with its tangled growth, and *Aconitum napellus*, rarely seen in such profusion elsewhere, its beautiful blue flowers being a most conspicuous object along the banks of the River Ely between Llantrisant and Cardiff.

The coastline of the County is very varied, and comprises muddy river estuaries, salt-marsh, lias cliffs, the Mountain Limestone cliffs of Gower and an extensive range of sand-hills, all of which deserve special mention.

The muddy estuaries of the Rivers Rhymney, Ely, Taff, and Tawe, produce *Cochlearia anglica*, *Trigonella ornithopodioides*, *Bupleurum tenuissimum*, *Apium graveolens*, *Aster*

Tripolium, var. *discoideus*, *Artemisia maritima*, *Glaux maritima*, *Atriplex littoralis*, *A. portulacoides*, *Suæda fruticosa*, *S. maritima*, *Juncus Gerardi*, *Triglochin maritimum*, and *Carex extensa*, while a large tract of tidal mud near the mouth of the Burry River is literally carpeted with *Limonium vulgare* and *Statice maritima*.

On the salt-marsh below Llanrhidian and also close to Llanmadoc, the great clumps of *Althæa officinalis* form a conspicuous feature in the landscape.

The sands and pebbly beaches afford *Glaucium flavum*, *Trifolium striatum*, *T. scabrum*, *Salsola Kali*, *Polygonum Raii*, and *Poa bulbosa*, whilst *Helleborus fœtidus*, *Astragalus glycyphyllos*, *Lathyrus sylvestris*, *Rubia peregrina*, and *Lithospernum purpureo-cœruleum* flourish in the undergrowth near by.

In the crevices of the stern Lias cliffs that fringe the coast from Penarth Head to Southerndown grow *Matthiola incana*, *Brassica oleracea*, *Inula crithmoides*, *Limonium binervosum*, *Adiantum Capillus-Veneris*, and *A. marinum*. Nash Point is famous for the rare thistle *Carduus tuberosus*, which is considered one of the most interesting plants in the County.

The sand-dunes, so conspicuous a feature of the Glamorganshire coast, produce many rare and beautiful plants. *Thalictrum dunense*, *Arabis hirsuta* sub-sp. *Retziana*, *Viola Curtisii*, *Geranium sanguineum*, *Filago minima*, *Senecio Jacobœ*, *Cynoglossum officinale*, *Echium vulgare*, *Euphorbia Paralias*, *E. portlandica*, and *Botrychium Lunaria* are among the first, after the advent of *Ammophila arenaria* to establish a footing on the shifting sand, while the slacks are often a tangled mass of *Rubus cæsius*, *Rosa spinosissima*, *Hydrocotyle vulgaris*, *Samolus Valerandi*, *Ligustrum vulgare*, *Blackstonia perfoliata*, *Centaurium pulchellum*, *Salix repens*, *Helleborine longifolia*, etc.

The rare *Limosella aquatica* var. *tenuifolia* grows on the sandy margins of Kenfig Pool.

Between Whitford Burrows and the muddy flats of the Burry River lies a rich belt of damp sandy soil which yields innumerable treasures for the field botanist, and in the early part of the year is a blaze of purplish-pink and yellow—*Ranunculus flammula*, *Hypericum quadrangulum*, *Listeria ovata*, *Iris Pseudacorus*, *Luchnis Flos-cuculi*, *Mentha sativa*, *Helleborine longifolia*, *Orchis incarnata*, *O. latifolia*, *O. maculata*, all take their turn during the months of June and July in helping to produce a picture of surpassing loveliness.

The Mountain Limestone of the Peninsula of Gower produces a flora differing considerably from that of other parts of the County. On the cliffs near Port Eynon and Horton and on the Worm's Head occur *Helianthemum canum*, *Potentilla verna*, *Calamintha Acinos*, *Asparagus maritimus*, and *Scilla verna*. On the walls of Penard Castle and on the cliffs near by, occur *Draba azoides*, the rarest plant in Glamorgan, which is to be found nowhere else in Britain, *Brassica monensis* (now almost extinct) *Hutchinsia petræa*, *Hippocrepis comosa*, and *Scabiosa columbaria*.

The limestone woods at Nicholaston and Oxwich contain many specimens of *Pyrus aria* and *Pyrus torminalis*, whilst among the shrubs and rocks between the woods and the beach occur *Hypericum montanum*, *Orobanche hederæ*, and a flourishing plantation of *Euphorbia Lathyrus* that has every appearance of being native and has been known to exist there for at least forty years.

The development of the coal industry, the erection of iron and steel works, and the building of docks at Cardiff, Penarth, Barry, Port Talbot, and Swansea, have produced many changes in the flora of Glamorganshire during the last 150 years. The fumes from the copper works at Swansea and Landore have had a disastrous effect on the vegetation for miles around, collieries and factories have polluted the water of many of the rivers, destroying the plants that once flourished on their banks, and the draining of Crumlin Bog has eliminated the habitats of many rare plants, among which may be mentioned *Drosera anglica* and *Andromeda polifolia*, once rather plentiful towards the northern extremity of the bog.

The spot near Cardiff where *Eryngium campestre* once grew is now occupied by the Roath Dock, while *Scirpus*

Holoschœnus was destroyed by the building of the Dowlais Works on the East Moors. *Convallaria majalis* has been exterminated, or very nearly so, from the Little Garth, and *Gagea lutea* has disappeared from the woods at Castell Coch. *Sibthorpia europœa* is said to have flourished in what is now one of the most populated parts of Cardiff, and *Orobanche elatior* has been searched for under the walls of Penrice Castle in vain. There are no recent records for *Ononis reclinata* (*Journal of Botany*, 1907, p. 280) or for *Matthiola sinuata*, which was once said to be plentiful on the sandhills between Swansea and the Mumbles, and *Euphorbia peplis* is probably extinct in the County. (*Journal of Botany*, 1907, p. 159).

What the flora has lost by building and the development of the coal and iron industries it has gained by the advent of railways and the opening up of commercial intercourse by means of steamers and sailing ships with different parts of the world. The ballast flora of the docks is naturally interesting, but foreign plants and others not suited to the soil usually flourish luxuriantly for a season or two and then disappear. Among adventive plants that have thoroughly established themselves in the County, North American species are by far the most prevalent, owing, no doubt, to the existence of similar climatic conditions, and *Compositae* is the natural order best represented. Many of these composite aliens have "invaded" waste ground near docks or railway sidings, and being of hardier growth have ousted nearly-allied forms and species not so well equipped to fight the battle for existence. This has been particularly noticeable in the case of the groundsels, but it is worth recording that the common indigenous type appears to be now re-asserting itself.

The forms of *Senecio vulgaris* occuring in the County require special mention. Dr. Trow, Principal of the University College of South Wales and Monmouthshire, has carried on experimental culture of the radiate forms at his garden at Penarth, and has recognised the following segregates as occurring in Glamorgan—(*a*) *prœcox* (Trow), (*b*) *erectus*

(*Trow*), (*c*) *erectus* variety *radiatus* (*Trow*), (*d*) *multicaulis* (*Trow*). He first examined the radiate type in 1891, and in 1894, Dr. C. T. Vachell called the attention of the members of the Cardiff Naturalists' Society to its appearance in a garden near Penarth Docks, and a specimen was sent to Kew for identification. It was named by Mr. Hemsley "*Senecio vulgaris* var. *radiatus*." *Senecio squalidus*, probably a native of Sicily, is now a common weed in waste places. It made its first appearance in the neighbourhood some years later, having probably spread along the embankment of the Great Western Railway from Swindon or Oxford. Its showy yellow flowers were much admired by the Marchioness of Bute, and its cultivation for a season or two in the Castle gardens no doubt assisted in its rapid naturalisation.

Matricaria suaveolens, a native of North America, is now fully established as a common weed in waste places about buildings and gardens and along the edges of roads. A few years ago it was found only in the neighbourhood of large towns. It is rapidly spreading into more rural situations. *Anaphalis margaritacea* extends along railway embankments and loose ground covering large tracts of country in the Rhymney Valley and elsewhere. It was mentioned as early as 1724 by Mr. Lhwyd (*Ray. Synopsis* ed. III, p. 182) as occurring "on the banks of the Rymney River for the space of at least 12 miles."

Impatiens fulva, a native of North America, is thoroughly naturalised at Llandough-juxta-Cowbridge, and is abundant in many parts of the Valley of the River Thaw from Llansannor to Gigman Bridge, and the pink Balsam, so well known in cottage gardens, is increasing rapidly along the banks of several of our rivers.

Mimulus Langsdorffii is specially conspicuous at Cheriton in Gower, *Lactuca scariola* established itself on the hurricane beach at Sully about the year 1900, and *Lactuca saligna* has appeared spasmodically about Llandough-juxta-Cardiff since 1899. *Brassica Cheiranthus* persists on the railway bank between Cardiff and Penarth as well as *Linaria repens* and the well-marked hybrid *Linaria repens* × *vulgaris.*

The soil and climate of the County assist the growth of introduced species of shrubs and trees, and the plantations of the Vale contain many fine examples of larch, spruce, and Weymouth pine. Myrtles, fuchsias, magnolias, and camellias flourish in the open, in warm situations, and rhododendrons are rarely seen in such profusion as at Penllergaer, the home of the eminent naturalist, Sir John Dillwyn Llewelyn.

It would be almost impossible not to say a word or two about the interesting flora of the Steep Holm, although the island is really included, both geologically and botanically, in the County of Somerset. It is the only wild station for *Pæonia corallina*, which was first added to the British Flora by T. B. Wright in 1803, and is also the home of *Allium Ampeloprasum*, first observed by Newton in 1689. The Island shows remains of habitations dating from very early times which may account for the presence of several of its rarer species of doubtful origin such as *Lavatera arborea* and *Smyrnium olusatrum*.

The Wyndcliffe, near Chepstow in Monmouthshire, is also a most interesting botanical hunting ground and is distant from Cardiff about 30 miles. The woods contain many rare plants, including *Pyrola minor*, *Pyrola secunda*, *Euphorbia stricta*, and *Cephalanthera longifolia*, while *Orobanche purpurea* is said to occur in a lane near Chepstow.

List of Books and Articles Referring to the Botany of Glamorganshire.

John Ray's Cat. Plant. Angl., 1670, Synopsis 3rd edition, 1724, and account of his Third Itinerary in 1662, edited in the "Memorials of John Ray" by Edwin Lankester.

The Rev. J. Lightfoot's Journal of a Botanical Excursion in Wales, 1773 (his collection of plants and descriptive MSS. are now in the British Museum). Printed in *Journal of Botany*, 1905.

Botanist's Guide through England and Wales by Turner and L. W. Dillwyn, 1805.

Rarer plants of Swansea, L. W. Dillwyn, 1828.

Outlines of the Geographical Distribution of British Plants, 1832, by H. C. Watson.

New Botanist's Guide, 1835, with Supplement, 1837.

Catalogue of Swansea Plants, *Mag. Nat. Hist.* 111, 1839.

Nicholson's Traveller's Cambrian Guide, 1840.

Topographical Botany, by H. C. Watson, 1874, Supplement in *Journal of Botany* by Arthur Bennett, F.L.S., 1905.

Geographical Distribution of British Plants, H. C. Watson, 1843.

Dr. J. W. G. Gutch's local list of plants near Swansea, in *Phytologist*, vol. 1, 1841 and 1842.

List supplementary to above by J. B. Flower and Edwin Lees, *Phyt.*, vol. 1, 1842.

List in *Phyt.*, vol. 1, 1843, by Thomas Westcombe.

List of plants in *Proceedings of Royal Institution of South Wales*,, 1844.

Materials for a Fauna and Flora of Swansea and neighbourhood, 1848. Not published.

Swansea plants in *Bot. Gazette*, 1849, by John Ball.

Phyt., 1850. Article by Joseph Woods.

Flora of Cardiff, by John Storrie. Published by the Cardiff Naturalists' Society, 1886.

British Association Handbook (Cardiff, 1891). Article by T. H. Thomas.

MSS. Notes prepared for 2nd edition of the Flora of Cardiff, by John Storrie.

The Flora of the Rhondda Valley, by Henry Harris, 1905.

A Flora of Glamorgan, by the Rev. J. Riddelsdell, 1907.

Further Glamorgan records by Rev. J. Riddelsdell, *Journal of Botany*, 1909.

The Flora of Glamorgan, edited by Prof. A. H. Trow, D.Sc., F.L.S., prepared under the direction of a Sub-Committee of the Cardiff Naturalists' Society, 1911.

References to Glamorganshire plants have appeared from time to time in the following journals and publications :—

Journal of Botany.

The Annual Reports of the Secretary of the Botanical Exchange Club and Society of the British Isles, edited by G. Claridge Druce, M.A., F.L.S.

The Transactions of the Cardiff Naturalists' Society.

The Proceedings of the Swansea Scientific Society, etc.

HERBARIA THAT MAY BE CONSULTED IN CARDIFF BY MEMBERS OF THE BRITISH ASSOCIATION.

Herbarium of the National Museum of Wales, 35, Park Place.

Herbarium belonging to Miss Vachell, 8, Cathedral Road, including Collection of the Rev. John Montgomery Traherne, born 1788.

Herbarium of the University College of South Wales and Monmouthshire, University College, Newport Road.

ZOOLOGY.

By T. W. PROGER, F.Z.S., and H. M. HALLETT, F.E.S.

SINCE the publication of the Handbook for the use of the British Association at its meeting in Cardiff in 1891, a good deal has been done towards working out the Fauna of the County of Glamorgan, but many orders are still awaiting their turn, and the following pages represent a sketch of those portions of the Fauna which have received any connected attention. The investigations are still being pursued, and it is hoped shortly to publish a volume dealing with the whole subject, and thus showing in a convenient form exactly what we know of the County Fauna.

The space at our disposal does not admit of lengthy lists of species being inserted, and we have therefore mentioned only those which are specially interesting or typical of the districts in which they occur.

General notes of the physical features of the County will be found in the article on Ornithology.

MAMMALIA.

Seven species of bats have been recorded in the County, the most interesting being the Greater and Lesser Horse-Shoe Bats and Daubenton's Bat, the first-named having been noted at the Mumbles and Coed-y-mwstwr, near Pencoed.

Insectivora are represented by three species of Shrews, the Hedgehog and Mole, both the latter being common.

The Wild Cat became extinct in the district about 1850.

The Fox is present in good numbers, finding considerable immunity in the lengthy expanse of sea cliffs.

The Mustelidae are represented by fewer species than of old, the Marten having disappeared within quite recent times, and the Polecat is on the verge of extinction. The Stoat and Weasel are common, and the Badger and Otter are to be found in fair numbers; an earth of the Badger is at present in occupation quite close to Penarth.

The Rodentia naturally furnish the largest number of species:—the Squirrel is common, the Dormouse not rare. The Black Rat, including the two forms known as the Alexandrine and Black Alexandrine occur fairly numerously about the Docks. The Field Vole, Bank Vole, and Water Vole are all plentiful, though we think the last-named is getting scarcer. The Hare remains only where preserved.

A full annotated list may be found in the "Transactions" of the Cardiff Naturalists' Society, Vol. 45, 1913.

ORNITHOLOGY.

Glamorgan is rich in its variety of bird life, a fact which is doubtless due to its varied physical features. To the north lies the hill district, where the higher ground is wild moorland broken up by many narrow valleys running more or less north and south. In the middle of the last century the sides of these valleys were thickly clothed with woods, and were the home of the Kite and the Buzzard. About this time the establishment of the iron industry, and later the exploitation of the coal measures, with the consequent great increase in the population, the felling of the woods for the manufacture of charcoal and pit props, caused many changes in the number and variety of our breeding species.

Most of the woodlands are now on the lower ground, nearer the coast, in the fertile Vale of Glamorgan, and in the peninsula of Gower. In the Vale (or "Bro Morganwg" as it is called in the "Booke of Glamorganshire's Anticquities," written by Rice Merrick in 1578), in several sheltered spots, summer migrants such as the Wheatear and Stonechat remain throughout the winter, and even the Willow Warblers remain with us until the middle of November.

The coastline of the County extends for upwards of 90 miles along the "Severn Sea," the cliffs in many places reach a height of over 100 feet, notably at Penarth Head, Porthkerry, Dunraven, and Worm's Head in Gower, and the

Peregrine and Raven still find nesting places not far from us. Choughs used to nest in certain places on the coast line, and we do not despair of their reappearance.

The Rivers of the County are the Taff, Ely, Ogmore, Neath, and Tawe ; the Rhymney forming the eastern boundary and the Llwchwr the western. All flow south into the Bristol Channel, and at their estuaries huge expanses of mud flats are uncovered at every tide, forming excellent feeding grounds for wildfowl and waders.

Kenfig Pool is our largest sheet of fresh water, 75 acres ; being situated near the sea and only separated from it by a high ridge of sand dunes, it is naturally a great resort of wildfowl, especially in winter, and this, with the Roath Park Lake (see guide book to Roath Park) and the City reservoirs at Llanishen, are great nesting places, the two last-named forming excellent sanctuaries, as they are never disturbed by gun fire. On the Morfa marshes, beyond Kenfig, White-fronted geese congregate in great numbers in winter. Crwmlyn Bog was in former times the nesting place of the Bittern, and the East Moors, Cardiff, another ancient resort, is still not infrequently visited by the species in winter.

Since the setting of the pole trap became illegal, Owls have become quite numerous in the County, and the Little Owl has firmly established itself.

We do not propose to give a full list of the birds of the County, but to mention the more interesting species which have occurred within the County boundaries. Those who are interested in the subject are referred to the "Birds of Glamorgan" prepared by a Committee appointed by the Cardiff Naturalists' Society, and published in Vol. 31 of the Society's Transactions, 1898-99, and the little Guide Book to the Roath Park and to the Ornithological Notes in the Transactions of the Society from 1900 onwards. These may be seen at the Public Reference Library in Trinity Street.

Of the 235 species recorded in the above lists, two, the Rusty Grackle (*Scolecophagus ferrugineus*) and the Little

CAROLINA CRAKE (*Porzana carolina*) are American birds, the latter appearing only once before, and the former for the first time in the British Lists.

The scientific arrangement and nomenclature adopted by Howard Saunders in his Manual of British Birds have been followed.

Since the publication of the Birds of Glamorgan in 1900, several species new to the County have been recorded, viz.:—the LONG-TAILED DUCK (*Harelda glacialis*), the BLACK-TAILED GODWIT (*Limosa belgica*), the EARED GREBE (*Podiceps nigricollis*), the GLOSSY IBIS (*Plegadis falcinellus*), the WHITE STORK (*Ciconia alba*), and the GREAT BUSTARD (*Otis tarda*).

RING OUSEL (*Turdus torquatus*). Summer visitor, breeds regularly in the hill districts.

ROCK THRUSH (*Monticola saxatilis*). A pair observed by the late Mr. R. Drane at Lavernock Point, August 23rd, 1868.

RUSTY GRACKLE (*Scolecophagus ferrugineus*). Shot on October 4th, 1881, on Grangetown Moors, Cardiff, new to Britain.

WHINCHAT (*Pratincola rubetra*). A few years ago this species was numerous ; it is now scarce.

REDSTART (*Ruticilla phoenicurus*). Nests regularly in a few places in the County.

BLACK REDSTART (*Ruticilla titys*). An occasional winter visitor.

NIGHTINGALE (*Daulias luscinia*). Not very numerous, breeds regularly and is increasing locally, becomes scarcer to the west of the County.

FIRE-CREST (*Regulus ignicapillus*). Not so uncommon in winter as it is supposed to be. A specimen in the Museum was obtained locally in March, 1915.

DIPPER (*Cinclus aquaticus*). Frequent on the mountain streams. One pair breeds regularly on the girder of a main line railway bridge within 4 miles of the City.

NUTHATCH (*Sitta caesia*). Rare. Frequently seen in the adjoining Counties of Monmouth and Brecon. The Hon. Odo Vivian records a Nuthatch picked up dead on a road near Swansea after the great storm in September, 1903.

White Wagtail (*Motacilla alba*). Frequently observed during the spring migration and has nested in this locality.

Golden Oriole (*Oriolus galbula*). Occasional summer visitor. Nested at Penarth in 1859, the four eggs mentioned by the late Mr. R. Drane as taken from this nest have now been brought together in the National Museum of Wales. A pair of these birds were seen at St. Hilary in 1878, and other pairs at Coedarhydyglyn in 1883 and 1886 by the late Captain George Traherne. He had reason to believe they nested on each occasion.

Waxwing (*Ampelis garrulus*). An occasional winter visitor. The latest specimen obtained is to be seen in the collection of Glamorgan Birds in the Museum in Trinity Street. Col. John I. D. Nicholl, of Merthyr Mawr, observed a Waxwing at a distance of 5 or 6 yards on the Porthcawl Golf Links in November, 1903. It had evidently just arrived.

Pied Flycatcher (*Musicapa atricapilla*). Occasional summer visitor. Although this species breeds regularly in the Counties to the north and north-west, it very rarely visits Glamorgan. The latest specimens obtained were one at Hirwain on May 1st, 1905, and another at Beaupré on February 15th, 1914.

Hawfinch (*Coccothraustes vulgaris*). Increasing in numbers, breeds regularly.

Goldfinch (*Carduelis elegans*). Now quite numerous, thanks to the limit put upon the bird-netters.

Siskin (*Chrysomitris spinus*). Small flocks visit us every winter, occasionally seen at Roath Park at this season, feeding on the alder seeds.

Tree Sparrow (*Passer montanus*). Not very common. feeding on the alder seeds.

Tree Sparrow (*Passer montanus*). Not very common. Local. Resident. Breeds regularly.

BRAMBLING (*Fringilla montifringilla*). Winter visitor, some times in large flocks in severe weather, but does not remain long after a thaw.

TWITE (*Ancanthis flavirostris*). Occasionally in wintry weather.

CROSSBILL (*Loxia curvirostra*). Winter visitor. There is reason to believe that this species has nested in the County. See note in "Birds of Glamorgan."

CORN BUNTING (*Emberiza miliaria*). Not very common. Local. Nests regularly.

CIRL BUNTING (*Emberiza cirlus*). Uncommon. Local. Breeds regularly.

SNOW BUNTING (*Plectrophenax nivalis*). Occasional winter visitor in small flocks. A pair of this species was found nesting on the Flat Holm in the Bristol Channel on May 20th, 1911. This nest, with the eggs and one of the birds, is preserved in the National Museum of Wales. This island is in the Parish of St. Mary, Cardiff.

ROSE-COLOURED STARLING (*Pastor roseus*). Has occurred five times in the County.

CHOUGH (*Pyrrhocorax graculus*). Rare. Has now disappeared as a breeding species in Glamorgan.

NUTCRACKER (*Nucifraga caryocatactes*). "I have seen one stuffed specimen in Wales, killed many years ago in the neighbouring County of Glamorgan" (Birds of Breconshire by E. Cambridge Phillips). The specimens in the Museum were shot at Pontypool, Mon., and Herefordshire.

RAVEN (*Corvus corax*). Increasing in numbers, nests within a few miles of the City.

HOODED CROW (*Corvus cornix*). Occasionally seen in winter.

WOODLARK (*Alauda arborea*). Not very common. Resident. Breeds regularly.

WRYNECK (*Iynx torquilla*). Not very common. Local. Breeds regularly.

HOOPOE (*Upupa epops*). Occasional summer visitor.

EAGLE OWL (*Bubo ignavus*). One was killed at Swansea in 1836. Dillwyn, "Flora and Fauna of Swansea."

Marsh Harrier (*Circus aeruginosus*). Has been observed three or four times. "Birds of Glamorgan."

Hen Harrier (*Circus cyaneus*). Rare. The specimen in the Museum was shot near Cowbridge, November 4th, 1895.

Common Buzzard (*Buteo vulgaris*). This species is increasing in numbers. It has recently nested within 20 miles of the City.

Kite (*Milvus ictinus*). Nested regularly in Leckwith Woods within sight of the City early in the last century. A Kite was caught in a trap set for Magpies in Redlands Wood at Cottrell, near Cardiff, on 12th April, 1917, and was presented to the Museum by Mrs. Mackintosh.

Honey Buzzard. (*Pernis apivorus*). Has appeared two or three times in the County. One of the specimens in the Museum was shot at Ruperra in June, 1876. A Honey Buzzard appeared in Porthkerry Park and remained during August, September, and part of October, 1913; we saw it many times when shooting there.

White-Tailed Eagle (*Haliaetus albicilla*). Visited the County in 1818, 1830, 1832, and was observed at Margam in 1860.

Peregrine Falcon (*Falco peregrinus*). Not very common. Breeds regularly within a few miles of the City. A male Peregrine frequented the tower of the City Hall during the autumn and winter months of 1917 and 1918. It destroyed numbers of homing pigeons and was eventually shot by an irate pigeon fancier.

Hobby (*Falco subbuteo*). Uncommon, breeds occasionally, Nest at Margam in 1897 and at Llanmaes near Llantwit Major, 1891. "During the summer of 1917 a small hawk, a Hobby, flew across the lawn surrounded by five or six swallows that were attacking it. When at a distance of only some 10 yards from me, the hawk struck and caught a swallow on the wing and carried it to a neighbouring tree." (Note sent by Mr. Hugh Bramwell, Pantyquesta, Pontyclun.)

RED-FOOTED FALCON (*Falco vespertinus*). A male bird of this species in very fine plumage was shot at St. Fagans on June 1st, 1903. Another bird of similar appearance, thought to be its mate, was seen near the same spot shortly afterwards.

MERLIN (*Falco aesalon*). Not very common, breeds in the hill district and on the sand dunes. We see it in the immediate locality during the spring and autumn migration.

OSPREY (*Pandion haliaetus*). Rare, occasional visitor. Has appeared three times in the County between 1860 and 1895, and Col. John I. D. Nicholl records one seen by him circling at a great height near the mouth of the River Ogmore on August 12th, 1904.

COMMON HERON (*Ardea cinerea*). There are three Heronries in the County, viz., at Margam, Penrice, and Hensol.

NIGHT HERON (*Nycticorax griseus*). Rare. A single occurrence in the spring of 1880 at Peterston, near Cardiff.

LITTLE BITTERN (*Ardetta minuta*). Rare. Several have been recorded, one of the latest was shot on Llangorse Lake, January 13th, 1907.

COMMON BITTERN (*Botaurus stellaris*). Occasionally in severe weather; several recorded between 1877 and 1916.

WHITE STORK (*Ciconia alba*). Three of these birds were observed in May, 1900, in a field near Bishopston in Gower. One of them was shot by a labourer and is now preserved in the National Museum of Wales.

GLOSSY IBIS (*Plegadis falcinellus*). An adult male was shot in a field on the bank of the Usk near Newport, Mon., on October 11th, 1902.

WHOOPER SWAN (*Cygnus musicus*). Occasional winter visitor.

BEWICK'S SWAN (*Cygnus bewicki*). Rare winter visitor. One shot at Taffs Well during the severe frost of 1895, now in the Museum.

GADWALL (*Anas strepera*). Occasional winter visitor. The latest occurrence was at Gelligaer, October 3rd, 1901.

SHOVELLER (*Spatula clypeata*). Not very common. Breeds regularly.

PINTAIL DUCK (*Dafila acuta*). Not very common. Winter visitor.

GARGANEY (*Querquedula circia*). Rare. Autumn visitor.

FERRUGINOUS DUCK (*Fuligula nyroca*). Rare; one record only. (Dillwyn, "Fauna and Flora of Swansea.")

LONG-TAILED DUCK (*Harelda glacialis*). An immature bird of this species frequented the upper and shallower end of the Roath Park Lake for some weeks in December, 1905, where it attracted attention by its remarkable diving powers. A pair of these birds frequented the Margam Marshes in December, 1911, one of which, the female, was shot and identified.

GOOSANDER (*Mergus merganser*). Winter visitor in small numbers to the river estuaries.

RED-BREASTED MERGANSER (*Mergus serrator*). Winter visitor in small numbers to the river estuaries.

SMEW (*Mergus albellus*). Not very common. Visits Roath Park Lake in severe weather.

PALLAS'S SAND GROUSE (*Syrrhaptes paradoxus*). Very rare; visited the County in 1888.

BLACK GROUSE (*Tetrao tetrix*). This species was once common in many parts of our hill district. A few wander over our border from Breconshire, and they may still nest very sparingly in the Hirwain district.

RED GROUSE (*Lagopus scoticus*). Few in numbers and decreasing. According to Dr. Washington David this species has been constant on the Llanwonno Hills for the past 30 years, and still occurs on the Gellyddu Mountain between the Aberdare and Merthyr Valleys.

QUAIL (*Coturnix communis*). Rare summer visitor. Has nested locally.

LITTLE CRAKE (*Porzana parva*). Rare. One recorded at Margam in 1839. (Dillwyn, "Flora and Fauna of Swansea.")

CAROLINA CRAKE (*Porzana carolina*). A single occurrence only in 1888.

GREAT BUSTARD (*Otis tarda*). A perfect specimen of this rare wanderer was shot by a farmer near Pontardawe on December 20th, 1902.

LITTLE BUSTARD (*Otis tetrax*). A single occurrence only. One shot at St. Athans in 1885.

REDSHANK (*Totanus calidris*). Very common in winter. A few pairs now breed regularly.

BLACK-TAILED GODWIT (*Limosa belgica*). Rare, only one record.

BLACK TERN (*Hydrochelidon nigra*). Occasional visitor, several recorded between the years 1888 and 1912.

WHITE-WINGED BLACK TERN (*Hydrochelidon leucoptera*). Occasional visitor, recorded two or three times.

LESSER TERN (*Sterna minuta*). Not very common. Now no longer breeds in the eastern part of the County.

LITTLE GULL (*Larus minutus*). Rare, occasional visitor.

BLACK-HEADED GULL (*Larus ridibundus*). Very common in winter. A few pairs have bred regularly in one locality for many years, but are now much disturbed.

COMMON GULL (*Larus canus*). Not very common. Winter visitor.

GLAUCOUS GULL (*Larus glaucus*). Uncommon winter visitor. The last specimen obtained on the East Moors, Cardiff, February 4th, 1903.

ICELAND GULL (*Larus leucopterus*). Rare. One shot off Penarth Head in January, 1898.

IVORY GULL (*Pagophila eburnea*). Rare. Two specimens only recorded.

GREAT SKUA (*Stercorarius catarrhactes*). Rare. Two or three times recorded. One shot at Sully now preserved in the National Museum of Wales.

POMATORHINE SKUA (*Stercorarius pomatorhinus*). Rare. The specimen in the Museum is said to have been shot by Stelfox, a fishermen, on the East Moors, Cardiff.

ARCTIC SKUA (*Stercorarius crepidatus*). Occasional visitor. generally immature birds.

BUFFON'S SKUA (*Stercorarius parasiticus*). Rare. A single record only.

LITTLE AUK (*Mergulus alle*). Several have been obtained after heavy gales at sea.

GREAT NORTHERN DIVER (*Colymbus glacialis*). Winter visitor to the Channel, and to Llanishen Reservoirs.

BLACK-THROATED DIVER (*Colymbus arcticus*). Rare. A single occurrence.

RED-THROATED DIVER (*Colymbus septentrionalis*). Rare, obtained two or three times between 1890 and 1910.

GREAT CRESTED GREBE (*Podiceps eristatus*). Breeds locally.

RED-NECKED GREBE (*Podiceps griseigena*). Several recorded.

BLACK-NECKED GREBE (*Podiceps nigricollis*). A single occurrence, October 28th, 1912.

SCLAVONIAN GREBE (*Podiceps auritus*). Not common, several recorded, see "Transactions," Cardiff Naturalists' Society, Vol. 37, 1904.

MANX SHEARWATER (*Puffinus anglorum*). Occasionally seen on the western side of the County in stormy weather; breeds in large numbers on the islands off the Pembrokeshire coast.

FORK-TAILED PETREL (*Oceanodroma leucorrhoa*). Rare; a single occurrence only.

STORM PETREL (*Procellaria pelagica*). Occasionally driven inland by storms.

FRESH WATER FISHES.

By H. EDGAR SALMON.

The County of Glamorgan has within its boundaries an exceptional number of rivers, streams, lakes, ponds, and reservoirs, and in them are to be found rather more than two-thirds of the fresh water fishes common to Great Britain. The Rivers Rhymney, Taff, Ely, and Ogmore were formerly noted for Salmon and Trout, but unfortunately at the present time, particularly in the lower reaches and estuaries, these fish are exceedingly scarce, owing to the pollution which has taken place from the development of collieries, coal washeries, tinplate and other works in the valleys through which these rivers run to the Bristol Channel.

Several reservoirs and colliery ponds are well stocked with trout; many from 2lbs. to 5lbs. in weight have been caught in the local reservoirs. In 1917 a trout weighing 7½lbs., and in 1919 two trout 7½lbs. and 10½lbs. respectively, were taken in a colliery pond at Hirwain.

The Sewin, or Western Sea-trout (*Salmo cambricus*), which is peculiar to Welsh waters is, according to the latest authority, a sea-run form of the common trout (*Salmo trutta*).

A Pike weighing 17¾lbs. from the lake at Hensol Castle is in the National Museum of Wales, and a number of pike from 12lbs. to 16¼lbs. have been caught in Kenfig Pool, a sheet of water 80 acres in extent, situated among the sand dunes near Porthcawl, on the coast. The Grayling was introduced into the River Ewenny in 1889 by Colonel Turbervill; they still survive in good numbers there. Eels are found in all waters, and when the elvers or young eels, commence to run, they are found in countless thousands all along the coast. The Grey Mullet is common in the docks at Cardiff, Penarth, Barry and Swansea, and in the estuaries. The more common fish are the Tench, Perch, Roach, Rudd, Gudgeon, Minnow, three spined Stickleback and Flounder; whilst among those less frequently taken are the Carp, Loach, Millers Thumb, Ten-spined Stickleback and the Lamprey or River Lamprey.

It is an extraordinary fact that there is no authentic record of either a Dace or a Chub having been caught in the County, although they are to be found in large numbers in the rivers of the adjoining County of Monmouth.

ENTOMOLOGY.

Hymenoptera-Aculeata.

This order has received a good deal of attention during the last few years, and a fair list, comprising some 250 species, has resulted. The weak part is the absence of any records from the high land at the north of the County, and no doubt many additions would be made if this district had been well worked.

With the large extent of sand dunes present, the records naturally show a good proportion of the Fossores and those Bees which affect this type of locality, and among these may be mentioned the usually rare *Methoca ichneumonides*, which has been taken in large numbers, both species of *Tiphia*, *Psammochares consobrinus*, *chalybeatus*, and *wesmaeli*, *Priocnemis affinis* commonly, *Ceratophorus morio*, once only; *Oxybelus mandibularis* and *argentatus*, the latter most abundantly: *Crabro styrius* and *tibialis*. *Colletes marginatus* is abundant at Porthcawl and Gower with other members of the genus, and accompanied by their parasites *Epeolus cruciger* and *notatus*. *Sphecodes pellucidus* is plentiful with its host *Andrena sericea*, *Halictus prasinus* plentiful but extremely local at Gower, *Andrena marginata* in fair numbers at Gower, and *A. hattorfiana* was taken many years ago near Swansea with its parasite *Nomada armata*; *N. obtusifrons* and *N. leucophthalma* are also recorded from Gower, but the former is much commoner inland at Taff's Well. *Coelioxys mandibularis* is very abundant on both sandhills, particularly at Porthcawl, as are several species of *Osmia*, the most interesting of which is *O. leucomelana* with its parasite *Stelis ornatula*.

The inland and woodland districts have provided records for *Sapyga 5-punctata*, *Psammochares nigerrimus*, *Psen dahlbomi*, *Crabro capitosus*, and others of this genus, whilst Cline Wood has provided a record for the very rare *C. signatus*. *Hylaeus dilatatus* turns up singly in several districts, the other members of this genus recorded are all of the commoner species. *Sphecodes* provides many species, of which *S. spinulosus* is very plentiful in one locality, *S. puncticeps*, *ferruginatus* and *hyalinatus* are not rare, and *S. rubicundus* has occurred once at Sully in a burrow of *Andrena flavipes*. *Halictus laevigatus* and *rufitarsus*, rarely, are the best of this genus, whilst the long list of *Andrena* is perhaps most remarkable for the presence of *A. bucephala* and the absence of *A. fulva*, the former occurring at Cwrt-yr-ala with its parasite *Nomada bucephalæ*.

Megachile versicolor has occurred once only, but probably has been overlooked ; *Osmia inermis* has occurred once at Bridgend. *Anthophora furcata* is by no means uncommon. All the species of *Psithyrus* have occurred, and most of the *Bombi*, the best of which are *B. lapponicus* and *soroensis*.

Of the 21 species of Ants which have been recorded, the most interesting are *Ponera punctatissima*, once only, the same number of appearances stands to the credit of *Myrmecina graminicola* and *Stenamma westwoodi*, whilst Sully is still the only known British habitat for *Myrmica schencki*. The most interesting species of *Formica* is *F. picea*, one example having been found in Gower by Mr. J. W. Allen.

Coleoptera.

By J. R. le B. TOMLIN, M.A., F.E.S

Glamorgan is the only Welsh County from which we have very early and withal reliable records. We know that Sir Joseph Banks collected on the coast near Swansea about 1785, and that excellent all-round naturalist Dillwyn has left us two admirable local lists, published in 1829 and 1848—the latter specially compiled for the 1848 meeting of the British Association at Swansea. A. R. Wallace was entomologising at this time in the Neath district, and Dillwyn's second list records some of his captures.

The County is undoubtedly rich in beetles and will eventually, I believe, be found to hold its own in this respect with the most productive ones of the South of England. Its most valuable asset is the long coastline with extensive salt-marshes and sand dunes—which exhibit a varied flors, and are watered here and there by streams or ponds, or with sandy bays, where stones and the flotsam of a great maritime highway provide abundant cover for littoral species.

The list is rich therefore in such genera as *Harpalus*, *Amara*, *Aphodius*, *Bembidium*, *Aleochara*, *Philonthus*, and *Bledius*, in Histeridæ and in some of the small Curculionidæ

Nebria complanata is often abundant in late summer and autumn, *Cicindela maritima* swarms at Oxwich Bay and elsewhere, *Elaphrus uliginosus* is not very rare too at Oxwich, and other noteworthy ground-beetles are *Dyschirius impunctipennis*, *Licinus depressus* (Llangenydd), *Panagaeus crux-major* (Llanmadoc), *Harpalus, attenuatus*, *H. melancholicus*, *Anchomenus micans* (Oxwich), *A. atratus; Bembidium assimile*, *B. normannum*, *B. doris*, *B. pallidipenne*, and *Tachypus pallipes*. The extremely rare *Amara fusca* was taken in numbers on the Crwmlyn Burrows over 60 years ago, and still awaits rediscovery Of Crwmlyn Bog one can only say " Ichabod " : most, if not all, of its treasures are extinct. Of other interesting beetles on the sand one may instance *Myrmecora brevipes*, *Homalium rugulipenne*, and the three species of *Phytosus*, which all occur under seaweed ; *Philonthus lepidus* and *P. puleus* (both in moss) ; ten species of *Bledius*, *Cyrtusa minuta* (once at Candleston) ; *Hypocaccus rugifrons* (under carrion) ; *Heterocerus femoralis* (Llanmadoc) ; *Psammobius sulcicollis ; Cardiophorus equiseti ; Psylliodes marcida* (on Cakile) ; *Helops pallida ; Cteniopus sulphureus ; Orthocerus muticus ; Nacerda melanura* (also a timber pest in Cardiff warehouses) ; *Hypera fasciculata ; Orthochaetes* (both species on violets, *O. insignis*, apparently confined to *V. curtisii*) ; *Bagous lutulosus ; Tychius squamulatus ; Ceuthorrhynchus pilosellus* (food plant still unknown), and *C. hirtulus*.

The salt-marshes harbour *Dryops algiricus* and *Aphodius plagiatus* in great profusion ; *Quedius hammianus*, *Syncalypta hirsuta ; Gronops ; Telephorus darwinianus* (common in June near mouth of River Ogmore) ; *Polydrosus chrysomela* and many others. The new British Halticid, *Crepidodera impressa* occurs sparingly at Oxwich on *Statice limonium*.

The high ground in the north of the County is inadequately explored, but has an extensive Alpine fauna. The wet moss of the mountain streams is prolific in the association of which *Quedius auricomus* and *Stenus guynemeri* may be taken as typical. Moles' nests and ants' nests yield the usual species, and the very rare *Claviger longicornis* has recently been taken at Sully with *Lasius umbratus*, and on the Little Garth.

The extensive felling of the past few years still gives unusual opportunities for wood work; thus a new British *Epuræa* (*E. distincta*) was taken in Gower last year; the very rare *Tomicus sexdentatus* has occurred in plenty in Clyne Woods under bark of *Pinus austriaca*, as well as *Trypodendron quercus* rarely, while *Hylesinus vittatus* is not uncommon on elm at Penarth.

It is not possible to single out more than a very few samples of general collecting:—*Bembidium testaceum*, *B. prasinum*, *B. punctulatum*, and many others of the genus are abundant in summer by the Taff; *Medon castaneus* (once at Southerndown); *Xantholinus fulgidus* common on the rubbish-tips at Danygraig; *Pæderus fuscipes* swarming in Oxwich Marsh; *Ancyrophorus aureus* in the River Perddyn and in a stream which enters Oxwich Bay; *Henoticus serratus* at Penarth; *Cathartus advena* in a Llandaff bakery; *Aphodius porcus* in several localities in early autumn; *A. quadrimaculatus* on the Garth; *Trichius fasciatus* scarce but generally distributed; *Malthodes* of which seven species occur at Llandaff; *Trigonogenius globulum* in a Cardiff mill; *Criocephalus ferus* in fir at Candleston; *Acanthocinus ædilis* frequently imported in timber; *Phyllotreta sinuata* (Llandaff and Crwmlyn Bog); *P. flexuosa* (Kolybion); *Cassida murræa* on *Inula dysenterica*; *C. chloris* (once at Candleston); *Pyrochroa coccinea* under bark at Castell Coch; *Mordella aculeata*; and *Mordellistena neuwaldeggiana*; *Choragus sheppardi* at Penarth; *Rhynchites uncinatus* on *Salix repens* at Candleston, and *Rhopalomesites* in Gower and at Sully.

Lepidoptera.

A full list of the Lepidoptera of the County has been published in the "Transactions" of the Cardiff Naturalists' Society, Vol. 50, 1917, to which those interested are referred. The list records some 600 of the Macro-Lepidoptera, but the weak part is the absence of records from the northern districts, and the almost entire absence of what are generally called the Micro-Lepidoptera. Among the more interesting records are those of *Papilio machaon* from the western boundary of the County ; *Aporia cratægi*, at one time abundant locally ; *Vanessa antiopa*, several records, the latest being at St. Nicholas in 1917 ; the alien *Anosia plexippus* has been captured on three occasions ; *Nomiades semiargus* (the Mazarine Blue) is one of the most interesting insects in the list, and probably Penarth was the locality for the last specimens taken in this country ; *Cerura biscuspis* has been taken at Llantrisant and Port Talbot ; *Palimpsestis fluctuosa*, a pair at Taff's Well in 1918 ; *Deiopeia pulchella*, once at Porthkerry, is one of the sporadic appearances which is the habit of this beautiful moth ; *Xylomiges conspicillaris* has occurred in Gower ; *Valeria oleagina* was taken on one occasion in the larval condition near Castell Coch by Mr. Williams, and the insect bred out. It is interesting to note that this insect was added to the British list by Donovan by a specimen taken at Fishguard, Pem., in 1800 ; *Xylophasia lateritia* taken at Porthkerry by Mr. W. E. R. Allen, is the only British record of this species. *Caradrina exigua* made its appearance in some numbers a few years ago, and many were taken at the electric light at Penarth ; *Dicycla oo*, once at Penarth by Mr. Howe ; *Lithophane furcifera* is another of the interesting species in our list, and was added to the British list by Mr. Evan John, it has been taken somewhat freely by him and Mr. H. W. Vivian, but we have not heard of any recent captures.

Plusia moneta was taken at Miskin by Mr. E. U. David very soon after its first appearance in this country. The Geometræ are very well represented, particularly the *Eupitheciæ*, some 36 species having been recorded ; the late Mr. H. W. Vivian paid special attention to this group. The scarce and sporadic *Sterrha sacraria* was taken in some numbers by Sir J. T. D. Llewelyn from 1869 to 1874. In the "Clearwings" the most notable are *S. andreniformis*, once near Cardiff in 1916, and *S. vespiformis*, once near Cardiff in 1894.

DIPTERA.

Our knowledge of this order is due almost entirely to Col. J. W. Yerbury, who has kindly given us a list of some 600 of the rarer species taken by him in the Gower and at Porthcawl. He writes that the list is fairly complete in some families and probably representative of the fauna of the County, but in others, *e.g.*, Syrphidæ, Anthomyidæ, Tachinidæ, etc., it is lamentably weak, he having almost entirely neglected these families. Among the more notable species are *Pamponerus germanicus*, rare on Newton Burrows; *Tachytrechus ripicola* at Porthcawl, added this species to the British list. Four specimens of the uncommon genus *Teucophorus* occurred near Nash Lighthouse; *Doros conopseus* was recorded by Capt. Bloomer (as *Ceria conopsoides*) in 1883; *Ptychoneura rufitarsis* was bred from the cells of the small wasp *Pemphredon lethifer*, and *Macronychia viatica* was added to the British list by Col. Yerbury from a specimen taken from the clutches of *Scatophaga stercoraria*.

It is impossible, in the space at our disposal, to go into details of the species recorded, but it is hoped to be able to publish this valuable list in the near future in the "Transactions" of the Cardiff Naturalists' Society.

HEMIPTERA.

Some 210 species of Heteroptera have been recorded for the County, and a full list was published in the "Transactions" of the Cardiff Naturalists' Society, Vol. 49, 1916, to which those interested are referred. The sand dunes produce such chracteristic species as *Pseudophloeus fallenii* and *Therapha hyoscyami* in numbers; *Chilacis typhæ* near Cardiff, *Dicyphus constrictus* at Oxwich Marsh, *Psallus vitellinus* at Sully, and *Corixa carinata* at Kenfig Pool are perhaps the most interesting species in the list.

ARANEIDEA.

Our records of the Spider fauna of the County are due entirely to the fact that Dr. Randall Jackson was resident at Ystrad, Rhondda, during the year 1901, and the results of his collecting were published in Vol. 39 of the "Transactions" of the Cardiff Naturalists' Society. His list comprises 177 species from only two localities, *i.e.*, Rhondda Valley and the Aberavon Sandhills, though only one day was spent in the latter district.

He found the fauna of the former locality distinctly northern in character, though southern forms such as *Evansia merens* and *Microneta cauta* occurred, and for both this was the second known locality, and in each case the female was new to science. Other spiders only previously found in one British locality were *Tetrilus arietinus, Sintula fausta, Porhomma miser* and *Troxochrus ignobilis.* Further rare and local forms were *Epeira sclopetaria, Xysticus ulmi* and *Metopobactrus prominulus,* whilst *Entelecara jacksonii* was new to science. At Aberavon *Synageles venator* and *Protadia patula* were known previously as British from Dorset and Sussex respectively, and *Styloctetor inuncans* was added to the British list.

LAND AND FRESH-WATER MOLLUSCA.

By J. DAVY DEAN.

The molluscan fauna of Glamorgan has been well worked, and as many as 113 species are recorded, practically all of which are represented in the Welsh Collection at the National Museum of Wales. Of these, 75 belong to the land and 38 to the fresh-water fauna.

Among the Helicidæ, *Helix pisana,* Müller, is undoubtedly the chief attraction to the collector. This local and strictly maritime species occurs in great abundance on the dunes at Porthcawl, and sparingly at Swansea. *Helix pisana* is a somewhat regressive species inhabiting south-western Europe, and is confined in these islands to Cornwall, South Wales, and the east coast of Ireland. *Helix nemoralis*, Linn., is abundant in Glamorgan, especially along the coast, dark forms predominating inland. *Helix hortensis*, Müller, is also common and has many interesting varieties, including the beautiful *arenicola*, McGill, remarkable for its frequency in the eastern districts. Two species, *Helicigona lapicida*, Linn., and *Theba cantiana*, Müller, reach their most westerly limit at Llantwit Major and Swansea. *Helicella* is represented by five species, all more or less discontinuous in their distribution. *Hygromia striolata*, Pfr., is everywhere abundant. *Hygromia fusca*, Mont., and *H. hispida*, L., are both local, while *Ashfordia granulata*, Alder, is confined to the western districts.

The more minute Celtic species are all present, with the exception of *Acanthinula lamellata*, Jeff.

The Zonitidæ are all represented, and although somewhat local in distribution are common where they occur. *Hyalinia lucida*, Drap., is abundant in the eastern district, and the white variety has been taken at Fonmon. *Hyalinia helvetica*, Blum., is common near the Brecon border and at Swansea, while *Zonitoides excavatus*, Bean, is characteristic of the coal measures and the oak forests inland.

Thirty-eight species of fresh-water Mollusca have been recorded and present an admixture of Celtic and Teutonic forms. All these are species of more or less general distribution and do not call for comment except in the case of *Physa*. There are three species, *Physa fontinalis*, Linn., *P. heterostropha*, Say, which have obtained a firm hold in the Cardiff area, and there is also a third and larger form, *Physa gyrina*, Say, which has not so far been recorded from any locality in Great Britain.

There are five species in the Vertiginidæ, all rare, three in the Clausiliidæ, inclusive of *Balea*, and both species of land operculates occur; *Cyclostoma* on the Lias cliffs of the coast, and *Acicula*, frequent in the oak-ash woodland of the limestone bordering the coal measures.

GEOLOGY OF THE CARDIFF DISTRICT.

By Professor ARTHUR HUBERT COX, D.Sc., Ph.D., F.G.S., University College of South Wales and Monmouthshire, Cardiff.

I. INTRODUCTION.

THE following account of the Geology of the Cardiff District is condensed from three papers, (i.) by the present writer,* (ii.) by Principal T. Franklin Sibly,† (iii.) by Dr. A. E. Trueman,‡ which deal with different portions of the geology of the District, and which were written for the recent visit (Easter, 1920) of the Geologists' Association.

The South Wales Coalfield and the areas immediately adjacent were resurveyed by the Officers of the Geological Survey between the years 1891 and 1914, and the geology of the various areas was described in the Survey Memoirs. The information contained in these Memoirs has been largely drawn upon in the preparation of this account. The list of those which deal with areas near to Cardiff is as follows :—

A. Strahan and others, "The Geology of the South Wales Coalfield," Parts I. Newport, II. Abergavenny, III. Cardiff, IV. Pontypridd, V. Merthyr Tydfil, VI. Bridgend, also "The Coals of South Wales," *Mem. Geol. Survey*, 1895-1915.

---

* "The Geology of the Cardiff District." Proc. Geol. Assoc., vol. xxxi. (1920), p. 45.

† "The Carboniferous Limestone of the Cardiff District," *ibid.*, p. 76.

‡ "The Liassic Rocks of the Cardiff District," *ibid*, p. 93.

The Cardiff Memoir, 2nd edition, 1912, contains a full Bibliography on the Geology of South Wales up to 1910.

Papers published subsequently to 1910 from which information has been drawn are :—

A. Strahan, "South Wales," *Geology in the Field*, 1910, vol. ii., p. 826.

F. J. North, "Note on the Silurian Inlier near Cardiff," *Geol. Mag.*, 1915, p. 385 ; "On a Boring . . . . . at Roath, Cardiff," *Trans. Cardiff Naturalists' Soc.*, vol. xlviii. (1915), p. 36.

H. K. Jordan, "The South Wales Coalfield, Part III.," *Proc. South Wales Inst. Eng.*, vol. xxxi. (1915).

F. Dixey and T. F. Sibly, "The Carboniferous Limestone Series on the South-Eastern Margin of the South Wales Coalfield," *Quart. Journ. Geol. Soc.*, vol. lxxiii. (1917), p. 111.

T. F. Sibly, "Special Reports on the Mineral Resources of Great Britain, vol. x., Iron Ores—The Hæmatites of the Forest of Dean and South Wales," *Mem. Geol. Survey*, 1919.

Definition of the Area.

The Cardiff District for the purpose of a geological description may be conveniently taken as the district within about 25 miles of Cardiff, in order to include the northern and southern crops of the eastern portion of the South Wales Coalfield. The greater part of this area is readily accessible by reason of the numerous railways that penetrate the Coalfield Valleys and centre on Cardiff or Newport. Such parts of Somersetshire as are within a 25-mile circle based on Cardiff are not considered.

II. The Stratigraphical Succession.

Within the area defined above the following formations are represented :—

Recent	Blown Sand, Alluvial Deposits, etc.
	unconformities.
Glacial	Fluvio-glacial deposits & Boulder Clay.
	great unconformity.

Mesozoic	Jurassic ..	.. Lower Lias.
	Triassic ..	.. Rhætic Keuper
	great unconformity.	
Carboniferous	Coal Measures ..	Upper or Llantwit and supra-Llantwit Measures. Pennant Grits. Lower or Steam-Coal Measures.
	Millstone Grit Series	
	local unconformable overstep.	
	Carboniferous Limestone Series	Main Limestone Lower Limestone Shales
Old Red Sandstone Series	.. ..	Quartz-conglomerates Brownstones Senni Beds Red Marls
Silurian	.. ..	Ludlow Series Wenlock Series
	base not seen.	

Excluding the thin superficial deposits, the history of the district is that of two long periods of deposition—one of Palæozoic age, and one of Mesozoic—separated by a period of powerful earth-movement with resultant uplift.

Deposition was practically continuous from Silurian to almost the close of Carboniferous time, so that the succession within these limits is almost unbroken. The detailed story of this period will be given in a later section of the paper. Towards the close of Carboniferous time there was a period of intense earth-movement (the Armorican movement) which threw the rocks into great folds and resulted in the elevation of the whole district above sea level. As a necessary cousequence great erosion took place, the tops of the folds were cut off, and a hill and valley system carved out. During this period the history of the district was entirely one of erosion and non-deposition, so that deposits of the period are unrepresented ; the Permian and a large part of the Trias (Bunter and portion of the Keuper) are absent. Eventually the lower ground was submerged beneath the waters of a great salt lake in which the Keuper Formation was deposited. Continued depression allowed the ocean waters access to the salt lake so that the Keuper is succeeded by the marine deposits of the Rhætic and Lower Lias. These various Mesozoic Formations—Keuper, Rhætic, and Lias—partly fill up hollows in the old hill and valley system, and rest with a strong unconformity on the older rocks below.

In all the vast interval from the time of the Armorican movement to the present day, no strong earth-waves have broken upon the district, so that Mesozoic rocks still remain almost horizontal. True there is some minor flexuring and faulting, or even local overfolding near the faults, but nothing to compare with the compressional structures among the Palæozoic rocks.

III. The General Structure.

It should be clear from the above outline account of the history of the district that the dominant structures are those that resulted from the Armorican folding. This folding had a general east to west trend as shown by the run of the structures. The whole area is dominated by the syncline of the South Wales Coalfield. From the synclinal area the Mesozoic cover has been almost completely stripped. South of the syncline is the Cardiff-Cowbridge anticline, an area underlain mainly by Carboniferous Limestone with Old Red Sandstone and even Silurian Rocks appearing in the core of the fold. Southward follow minor synclinal and anticlinal rolls, one of the latter bringing up Old Red Sandstone in Barry Harbour. The main anticlinal areas formed low ground in Triassic times just as now, so that Mesozoic rocks were laid down and are still preserved over a wide extent, concealing the structures of the Palæozoic rocks below (fig. 4).

The two principal folds are not, however, quite simple structures. For example, within the coalfield synclinal area there is a strong anticline—the Pontypridd Anticline—of great tectonic and industrial importance (see p. 00). Then in the Old Red Sandstone area that forms the middle limb between the Coalfield Syncline and the Cardiff-Cowbridge Anticline there are several minor folds which strike parallel with the main folds,but which cannot as a rule be followed for any great distance. Examples are seen on Craig Llanishen (fig. 7), and also at Tongwynlais, where the outcrop of the Lower Limestone Shales is duplicated by an anti-clinal roll. Other minor folds are found among the Silurian rocks in the core of the Cardiff anticline (fig. 1). The pitch of all these folds in East Glamorgan is to the west.

IV. Scenery of the District.

Scenically there are two markedly contrasting types within the district: (i.) the high ground of the Coalfield and the Coalfield Rim in the North, (ii.) the lower ground of the Vale of Glamorgan to the South. In the Coalfield the dominating feature is the deeply trenched upland carved out of the Pennant Grits—an area of gently swelling or sometimes flat-topped hills, almost uncultivated and uninhabited, and given over to rough pasturage. Wherever synclines bring in the clays of the higher measures, or anticlines bring up those of the Lower or Steam Coal Measures, the Pennant upland gives place to hollows bounded in the one case by gentle dip slopes, or in the other by steep scarp slopes (fig. 2). Where not occupied for industrial purposes the ground in each depression is seen to be more fertile than that of the uplands to such a degree that the geological boundaries are readily defined. Apart from such often wide depressions the upland is gashed by deeply-cut valleys which afford a home to a dense population, often excessively overcrowded owing to the steepness and height of the valley sides. The overcrowding is the more noticeable by reason of the proximity of the uninhabited hills. South and east of the Coalfield proper comes the Coalfield Rim, a narrow belt of high ground with the parallel scarps of the Carboniferous Limestone and the Quartz-conglomerates of the Old Red Sandstone (fig. 7). In the Ely valley and the district west thereof the elevated rim is absent, as the barrier was broken down in Triassic times, so that Mesozoic deposits rest directly on the Coal Measures.

South of the elevated tract of the Coalfield lies the lower fertile, agricultural area of the Vale of Glamorgan, partly drift-covered, and floored mainly by the Triassic and Liassic marls or clays, through which appear here and there the underlying Carboniferous Limestone and Old Red Sandstone. The presence of these patches, and also the presence of escarpments due to the Rhætic Beds and the Limestones (*Ostrea*-Beds) of the Lower Lias, serve to diversify the scenery of this tract, while the unusual percentage of limestones in both the Trias and the Lias mitigates the clayey nature of the soil. The average elevation of this tract is rather over 200 feet, and it terminates seawards in a long line of cliffs. It has already been pointed out that the lower elevation of this tract, situate on the Cardiff-Cowbridge anticline, dates even from Triassic time, and is accounted for by the well-known principle that anticlinal areas give way under denudation more rapidly than synclinal areas. Thus we find that the Carboniferous Limestones patches, even when fairly extensive, do not form ground that rises more than 100 feet, or perhaps 200 feet, above the general level of the Vale, so that the Limestone in the Vale behaves very differently to that in the Coalfield Rim, where it forms a marked scarp. The effect of the present-day denudation in the Vale is to clear away the soft Mesozoic cover and to restore the Triassic outlines.

Eastwards the low ground of the Vale passes into the low and undulating plain formed by the Red Marls of the Old Red Sandstone Series, and this plain sinks gradually southwards to the Bristol Channel. Along this stretch the coast is bordered by a wide stretch of alluvium, partly brought down by the rivers, partly deposited by the strong tides.

Further details of the scenic effects of each Formation will be found below under the descriptions of the individual Formations.

Near Cardiff there appear to be the relics of a ? Pliocene plain about 200 feet above present sea level, for the Liassic Rocks of Leckwith and Penarth, and the Silurian Rocks of Penylan Hill and Rumney are all planed off at about the same height.

The River System.

The river system gives a good example of superposed drainage which is in the main independent of geological structure, but which is endeavouring to adjust itself as far as possible to that structure. There are three main gathering grounds—two just north of and the third within the Coalfield. Those north of the Coalfield are the Millstone Grit dip slope, 1,800 feet above O.D., north of Tredegar, from which flow the Ebbw, Sirhowy, and Rhymney Rivers, and the dip slope of the Brownstones of the Brecon Beacons, 2,000 to 2,900 feet O.D., from which flow the Taff and the streams that drain into Glyn Neath. The third centre is the high ground 1,800-1,900 feet O.D. made by the Pennant scarp on Craig-y-Llyn and hills to the south, whence flow the Rhondda and Ogmore Rivers, and others. Most of the streams mentioned were initiated on southward or south-south-eastwards facing dip slopes, and they maintain a south-south-east course right across the Coalfield, flowing in deep steep-sided valleys which cross the strike of the various beds. In their lower and middle parts the streams have nearly reached base-level and the valley bottoms are filled with alluvial deposits. The more rapid cutting of some of the streams has led to several cases of river capture, of which one example may be mentioned, the capture of the Bargoed Taff by the Taff. The Bargoed Taff originally went through a valley (wind-gap) past Nelson to Ystrad Mynach as a tributary to the Rhymney until captured by the faster cutting Taff.

The streams have breached the south rim of the Coalfield in a series of gorges which afford lines of communication to the Coalfield, and so have determined the positions of the chief ports; Cardiff was determined by the Taff gorge, Newport by the Ebbw. The winding course taken by the Rhymney below Caerphilly in its attempt to escape from the Coalfield is noteworthy. It is apparently determined in part by the presence of the synclinal basin of soft Upper Coal Measures which the river follows as long as possible.

A feature of considerable interest is the river capture and rejuvenation at the head of the Vale of Neath, due to the rapid cutting back of the River Neath along the crush zone of the north-east—south-west disturbance. This has led to the capture of several of the rivers—the Perddyn, the Upper

Neath or Nedd, Mellte, and Hepste—that flow southwards from the dip slope of the Brecon Beacons and that formerly ran via Hirwain and Aberdare to the Taff Valley. It is noticeable that the valley through Hirwain and Aberdare is aligned with the Perddyn Valley, and that the Hepste probably entered this valley through the wind-gap at Penderyn. The capture has resulted in rejuvenation, so that these streams have cup deep, and sometimes almost inaccessible gorges through the Millstone Grit moorlands or through the underlying Carboniferous Limestone. Further interesting features in this district are the numerous swallow-holes, and the underground courses of several of the streams due to the ready solubility of the Carboniferous Limestone. Most of the streams flow underground in parts of their courses. Perhaps the best example is the Mellte at Porth-yr-Ogof near Ystradfellte, where the stream flows towards and into a great limestone cliff by way of a cave which can be penetrated on foot for some distance. The river emerges after an underground course of 220 yards. The old course, now dry, can be seen on top of the cliff about 100 feet above the present water level.

Glacial diversions occur here and there ; an example near Cardiff is the small stream that drains a valley near Wenvoe, excavated in soft Keuper Marls between two inliers of Carboniferous Limestone. The old Keuper valley continues towards Cadoxton and so to Barry, but the depression has been blocked and the stream has accordingly been forced to cut a gorge through the Carboniferous Inlier of Cwrt-yr-ala.

V. Geological History of the District.

Evidence of continued Instability and Proximity of Shore-lines.

Apart from the industrial importance of its geological structure one of the most noticeable features is the pronounced lateral variation that takes place in the lithology and succession of practically every Formation within the 25 mile limits of the district. This is true of the Carboniferous Limestone, Millstone Grit, Coal Measures, Trias, Rhætic, and Lias, and it naturally imparts additional variety to the geology of the area. Changes so marked and so persistent throughout the vertical scale invite further investigation. For those desirous of obtaining a general view it will be convenient to outline the chief events of the period before proceeding to a detailed description of the rocks themselves, although this procedure is, of course, a reversal of the method followed by the geologist, who must deduce the geological history of any district after he has made a detailed study of its rocks.

Examining each Formation in turn we obtain evidence that throughout long geological periods the district has lain in the immediate vicinity of a shore-line, which, owing to oscillation of earth-movements, was situated sometimes to the north and sometimes to the south of the Cardiff area.

The known geological history opens with marine conditions prevailing, under which the earliest visible rocks—the Silurian—were deposited. The Silurian rocks of the Cardiff District represent a rather sandy and shallow-water facies of the Shropshire type, and they indicate that the shore-line could not have been far distant. From the known distribution of the Silurian Rocks of Wales it is clear that this shore-line must have been to the south, though evidence as to its exact position is wanting.

At the close of Silurian times a reversal of conditions took place, brought about by the post-Silurian or "Caledonian" mountain-building movements which affected chiefly the districts further north, such as North Wales, the Lake District, and Scotland, but made their influences felt as far south as the Cardiff district. The Old Red Sandstone of South Wales was laid down in great inland fresh-water lakes in contrast to the homo taxial marine deposits of Devonshire and Cornwall. The shore-line of the Middle Devonian sea must therefore have been situate between Devonshire and South Wales.

In early Carboniferous time a gradual but prolonged subsidence affected a wide area so that the shore-line receded northwards, the sea gained access to the former fresh-water lakes, and the Carboniferous Limestone was deposited. Comparison of the Carboniferous Limestone of South Wales and Somersetshire (the "South-Western Province") with that of North Wales and Derbyshire shows that a land barrier was present across the English Midlands so that the shore-line of the South-Western Province sea had retreated as far north as the Clee Hills. But we have evidence that the depression was not uniform owing to instability along a line passing somewhere east of Cardiff. Along this line the regional depression was checked, or even perhaps partly counteracted by local uplift, which rendered the Carboniferous Limestone sea shallower to the north-east of Cardiff than to the south, and thus led to great differences in the lithology of the sediments deposited in the two areas, so that the Carboniferous Limestone of the South Crop differs considerably from that further south in the Vale of Glamorgan and Barry districts.

Towards the close of Lower Carboniferous times there was a reversal of the main movement of depression; elevation took place, ushering in the shallow-water Millstone Grit conditions. With this general uplift taking place, the local uplift to the east became more pronounced, leading to actual denudation of the Carboniferous Limestone, so that on the east side of the Coalfield there is an unconformity between the Millstone Grit and the Limestone; the upper portion of the Limestone is missing and also the lower portion of the Millstone Grit.

The further history is then one of gentle subsidence, approximately keeping pace with sedimentation so as to maintain comparatively shallow-water conditions for long periods, during which the Coal Measures were deposited. But the regional depression suffered several checks, leading to the production of alternations of shallow and deeper-water beds, and perhaps even to local overlap towards the east, and to unconformability in the Forest of Dean* and

* T. F. Sibly "Geological Structure of the Forest of Dean," *Geol. Mag.* 1918, p. 25.

Bristol Coalfields,* so that once again shore-lines were established within a few miles of the Cardiff District.

Still later there followed the great Armorican uplift and folding mentioned above (p. 251) and the shore-line was pushed far to the south. But with renewed depression shore-lines were once again established close to Cardiff, as the land gradually subsided beneath the waters of the Triassic salt-lake. Subsequently continued depression let in the waters of the Rhætic and Liassic seas and the shore-line was pushed northwards towards the high ground of the present Coalfield.

The long interval of time from the Lower Liassic to the Glacial Periods is a blank so far as the Cardiff District is concerned, as no deposits belonging to that long interval have survived. At some period subsequent to the Lower Liassic elevation took place, resulting in renewed erosion, which has continued almost without intermission to the present day. Once more we find a coast-line established in the district, and the submerged forests of South Wales, and the nature of the post-Glacial deposits of Barry and elsewhere, prove the sea-level to have fluctuated within recent times.

It is noteworthy that despite this continued instability throughout the geological ages and the great changes that have taken place in consequence, there are no igneous rocks within the district. The nearest are the Lower Carboniferous lavas of Weston-super-Mare, Somersetshire, and the solitary intrusion of monchiquite in the Old Red Sandstone, towards Usk.†

VI. Description of the Formations.

A. Palæozoic. 1. Silurian.

The Silurian are the oldest rocks exposed in the district, and are brought up as an inlier along the core of the Cardiff-Cowbridge anticline. Their outcrop occupies only a small area, two or three square miles, intersected by the Rhymney

*F. Dixey "Relation of the Coal Measures to the Lower Carboniferous Rocks in the Clapton-Clevedon District, Somerset," *ibid.* 1915, p. 312.

† W. S. Boulton, *Q.J.G.S.*, vol lxvii (1911), p. 460.

Valley. Owing to the superior hardness of the strata the outcrop gives rise to small hills—Penylan Hill—which stands up some 200 feet above the neighbouring country. The rocks do not lend themselves to accurate investigation since the base of the series is not reached, and much of the outcrop is buried beneath Keuper Marls and alluvial deposits, which lap round the south sides of the hills (fig. 1).* Further, the junction with the overlying Old Red Sandstone is situate in low ground, almost destitute of exposures. The total thickness exposed is about 1,000 feet.

As already remarked, the rocks present a sandy variant of the Shropshire facies, quite distinct from the graptolitic facies of other Welsh areas. The beds were evidently deposited in shallow-water at no great distance from a coast-line. They are essentially grits, sandstones often micaceous, and sandy shales almost destitute of limestone bands, and even such bands as originally contained some calcareous material are now largely decalcified. The essentially local lithology, the absence of limestone bands, and the long-time ranges of the commonest fossils, the brachiopods, render difficult any exact correlation with the type district of Shropshire. As might be expected in such arenaceous rocks the principal fossils are small brachiopods, together with some trilobites. Corals, and still more, graptolites are of rare occurrence.

The occurrence of such fossils as ***Pentamerus oblongus*** suggests that Llandovery beds are represented. Alternating greenish-grey shales and sandstones, some flaggy, others highly concretionary, represent the Wenlock. Intercalated in this series is the Rumney Grit, the most distinctive band of the local Silurian. It consists of massive and flaggy sandstones, blue-grey when fresh, but weathering yellow. Some of the sandstones show ripple marking and contain abundant "fucoids." This grit makes a small feature, which can be followed across the inlier. It is exposed in quarries near Rumney (300 yards E.N.E. of Rhymney

* The illustrations will be found at the end of the article.

Bridge), where the beds yield some lamellibranchs, *Grammysia cingulata*, *Ctenodonta*, also the problematical seed-like bodies, *Pachytheca*. The flaggy grits exposed among the Wenlock Shales at Roath Park are usually stated to be on the same horizon as the Rhymney Grit. The Ludlow is represented by alternations of thin flaggy sandstones, grey and blue micaceous mudstones and shales, etc., with various lamellibranchs and gasteropods, but the beds are at present hardly exposed, so that the position of the junction with the overlying Old Red Sandstone and the nature of the beds at the junction are largely matters of inference ; so far as can be made out, the Old Red Sandstone follows on conformably and its base is usually taken at a convenient thin grit band, which is locally a *bone-bed*, containing fish-teeth and scales, as in the east bank of the Rhymney. Colour is no criterion as a result of iron staining through percolation from the former overlying Trias cover. If the relationship between the two systems is a truly conformable one, and the lower limit of the Old Red has been placed correctly, the Downtonian Beds must be much thinner here than in Shropshire. Any future excavations for building or other purposes should be carefully watched in order that such information as becomes obtainable should not be lost. It is especially desirable that the exact relationship of the Silurian to the Old Red Sandstone should be determined in view of the recent work on the Downtonian of Shropshire* and South Staffordshire.†

Fifteen miles to the north-east is another inlier of Silurian—the Usk inlier—brought up by a dome-shaped fold. The Silurian Rocks of this inlier have been recently described‡ ; they are essentially of Shropshire facies and do not call for further notice here.

2. OLD RED SANDSTONE.

The Old Red Sandstone, which crops out almost the whole way round the Coalfield Syncline, is a continuation

* L. D. Stamp, "The Highest Silurian Rocks of Clun Forest," *Q.J.G.S.*, vol. lxxiv. (1918), p. 221.

† W. Wickham King & W. J. Lewis, *Proc. B'ham Nat. Hist. and Phil. Soc.* vol. xiv. (1917), p. 90

‡ C. I. Gardiner, *Proc. Cotteswold Nat. F. Club*, vol. xix. (1916), p. 129.

of the extensive Herefordshire outcrop, and the strata are entirely of fresh-water facies as contrasted with the marine type prevailing across the Channel from Minehead to Ilfracombe. The series has been divided by the Officers of the Geological Survey into four unequal divisions :—

Grey Grits and Conglomerates	Grey and brown quartz-conglomerates and pebbly quartzites overlain by yellow and red marls and sandstones.
Brownstones	Rapid alternations of red and brown sand stones, shales and marls, with some pebbly horizons.
Senni Beds	Alternations of green, dull-red, and chocolate sandstones, shales and marls, with numerous thin, green and red, conglomeratic cornstones.
Red Marls	Red marls with red micaceous sandstone and thin cornstones, also some conglomerate bands.

Along the western margin of the Coalfield in the Abergavenny, Newport, and Cardiff area, the Senni Beds have not been separated from the Brownstones. In the immediate neighbourhood of Cardiff the thickness is estimated at 2,800 feet for the Red Marls, 500 feet for the Brownstones, and 200 feet for the conglomerates. These estimates are obtained from the section along the Rhymney Railway from Roath Park to Cefn On. The strata thicken considerably towards the north, the Brownstones, for instance, are about 1,200 feet in the Abergavenny district and nearly 1,500 feet on the Brecon Beacons, while the Red Marls thicken to 4,500 feet in the Ammanford district.

The relation of the Old Red Sandstone to the Silurian is as already stated not very clear in the Cardiff district, but so far as can be seen at present, it compares generally with that in Shropshire, that is, there is practically a transition, not a strong unconformity such as is found west of Carmarthen.

The passage to the Carboniferous Limestone above is also one of apparently perfect conformity. Within the Old Red Series itself there is no evidence of any real stratigraphical break, despite the frequent bands of cornstones and conglomerates. Small breaks due to contemporaneous erosion by powerful currents are common enough, but are purely local in character and of no real significance for purposes of correlation.

Fossils are extremely scarce throughout the series. Fish remains, *Pteraspis*, *Cephalaspis*, and others have been obtained at several localities round Newport (Gold Tops, Stow Hill Road, Maindee), and from a cutting on the Taff Vale mineral railway at Croft-y-genau, 2 miles west of the Taff at Radyr; also plant remains near Llandegorth from beds quite low in the series, near the Usk Silurian inlier.

The Upper Old Red Sandstone age of the higher beds is proved by the occurrence of *Archanodon jukesi* near Talgarth, 5½ miles east of Caerleon, and of *Holoptychius* near Abergavenny, also by plant remains, *Stigmaria ficoides*, and others, near Kidwelly, outside the Cardiff district proper.

As regards scenery, the Quartz-conglomerates with the underlying Brownstones give rise to a scarp, which faces outwards from the Coalfield and which, in conjunction with the neighbouring Carboniferous Limestone scarp, serves to shut off Cardiff from the mining districts; communication is restricted to the few valleys, Taff, Rhymney, Sirhowy, etc., that breach the barrier. West of the Taff the escarpment rapidly sinks beneath the unconformable Keuper cover. From the foot of the Brownstone scarp the area occupied by the underlying Red Marls stretches away as a wide, much dissected plain, the surface of which is varied by innumerable little ridges and swells caused by outcrops of the intercalated sandstone and cornstone bands. In the Cardiff area this plain falls gradually seawards, and the Marl finally sinks southwards beneath the alluvium that borders the Bristol Channel, or south-westwards beneath the unconformable cover of Keuper Marl. North-east of Cardiff town the plain is interrupted by the Silurian dome that forms Penylan Hill. The area covered by these Red Marls is exceedingly fertile, hence the origin of the term *cornstone* applied to the calcareous bands. As in Herefordshire the soil seems specially suited to orchards, and the verdant character of the country contrasts strongly with the bare moorlands made by the Millstone Grit and Pennant Grit, and with the slatey areas occupied by the Lower Palæozoic rocks of Wales.

Near Cardiff, on Craig Llanishen and other hills, the Brownstone scarp averages 700-800 feet in height. Followed northward along the eastern margin of the Coalfield, the escarpment increases in height as the Brownstones thicken to the north and north-west. Thus on the Sugar Loaf, near Abergavenny, it attains a height of 1,950 feet, in the Black Mountains 2,600 feet, and culminates on the Brecon Beacons 2,900 feet. North of the Coalfield, where dips are much less than in the south, the dip slope is correspondingly long and gentle, consequently the scarp is further removed from the mining areas. Thus along the line of the Rhymney Railway between Cardiff and Caerphilly the Brownstone scarp is only ½ mile distant from the Lower Coal Measure outcrop (figs. 2 and 6), whereas north of the Coalfield the outcrops of corresponding beds are 10 miles apart.

3. Carboniferous.

The Carboniferous is both structurally and, of course, economically the most prominent system in South Wales. As usual in South Britain the lower division (Carboniferous Limestone) is a marine formation, the upper division (Millstone Grit and Coal Measures) is of shallow-water marine, estuarine, or fresh-water origin. The lowest rocks follow conformably upon the Old Red Sandstone, but between the upper and the lower divisions, that is, between the Millstone Grit and the Carboniferous Limestone, there is local overstep and unconformity, probably also breaks and small overlaps at higher horizons in the Coal Measures. The top of the System is nowhere reached in South Wales, owing to the great unconformity that brings on Mesozoic Rocks.

One of the most noticeable features is the eastward thinning that affects every division of the Carboniferous System as developed in South Wales. This is evidenced by the following figures :—

	Swansea district.	Taff Valley (South Crop).	Pontypool district.
Upper Coal Measures			
Pennant Series	2400	1600	650
Lower Coal Measures	3000	1000-1700	800
Farewell Rock	18C	abt. 450	400
Millstone Grit Shales	1000		
Basal Grits	350	*absent*	*overstep*
Carboniferous Limestone ..	3000	2000	350

This eastward thinning is due to difference in sedimentation and is helped by the breaks mentioned above ; the Millstone Grit overstep and unconformity for example only develop to the east.

(*i.*) *The Carboniferous Limestone.*

The marine beds of the Carboniferous Limestone follow quite conformably upon the fresh-water beds of the Old Red Sandstone. Upwards the Formation is unconformably overstepped by the Millstone Grit Series. The Limestone forms a strip right around the eastern half of the Coalfield—the North and South Crops—and it recurs in the Vale of Glamorgan, where it underlies most of the area affected by the Cardiff-Cowbridge Anticline and the folds south thereof (fig. 2). In the latter area the rocks only appear at intervals from beneath the unconformable cover of Mesozoic Rocks.

The Limestone of the Cardiff District belongs to the area or deposition known as the "South-Western Province," which embraces the whole of South Wales together with Gloucestershire and Somersetshire, in which the section in the Avon Gorge, near Bristol, may be regarded as the type section. Conditions of deposition were not, however, uniform over the whole area of the South-Western Province, and notably in the Cardiff District they were subject to variation. This caused considerable lateral change in the Formation, both in a north-south and in an east-west direction. Consequently we now find marked differences between the Limestone of the Coalfield Rim and that of the Vale of Glamorgan, and yet again between the developments in the North and South Crops. The Limestone of the Vale of Glamorgan compares most closely with the Somersetshire development, as might be expected in view of the proximity of the two areas. It may be noted here that very little modern zonal work has been accomplished in the Vale of Glamorgan, so that a rich field still remains open to the investigator. Concerning the Limestone of the South Crop we have fuller information, thanks to the detailed work of Mr. F. Dixey and Principal Sibly. The following is quoted *in extenso* from Principal Sibly's most recent paper*:—

"1. CENTRAL AREA.

"A survey of the Carboniferous Limestone is best commenced in that part of the district in which the sequence of zones is most nearly complete. This is the case in a central tract, extending from the Ewenny valley to the neighbourhood of Creigiau, where the limestone-outcrop belongs equally to the south-eastern rim of the Coalfield basin and the northern limb of the Cardiff-Cowbridge Anticline.

"The succession in this part of the district is illustrated by the Ruthin-Llansannor and Miskin vertical sections in fig. 4, and also by the Miskin horizontal section, fig. 5. The total thickness of the Carboniferous Limestone attains a maximum of some 2,700 feet in the west and falls to about 2,100 feet in the east, the eastward diminution being due to the loss of the *Dibunophyllum* beds by the overstep of the Millstone Grit. The sequence may be summarised as follows :—

	Zone		Feet.
Main Limestone	D { $?D_2$ (part of)	Limestones, including pseudo-breccias..	600
	D_1, S_2	*Modiola* phase, characterised especially by limestones with pisolitic structures; *Seminula* Oolite	600
	C_2+S_1	Limestones, including much oolite; *Modiola* phase	350
	C_1	*Caninia* Oolite .. ..; *Laminosa* Dolomite .. .. ..; (γ C_1) Crinoidal limestones	550
	Z	Crinoidal limestones and dolomites. Chert near the base	300
Lower Limestone Shales	K	Shales with thin limestones .. ..; Crinoidal limestone and oolite .. ..; Limestones and shales	300
		Total	2700

* "The Carboniferous Limestone of the Cardiff District," *Proc. Geol. Assoc.* vol. xxxi (1920), p. 81.

"The Lower Limestone Shales, in so far as they are exposed, present a normal facies of limestones and shales. The *Modiola* phase of K, which is well seen further east on the Coalfield margin, is not exposed.

"The Main Limestone is composed principally of undolomitized or little dolomitized limestones (including oolites) with a standard fauna of crinoids, corals, and brachiopods. But contemporaneous dolomitization, with impoverishment of the fauna, is important at two levels, namely: (1) in Z, and (2) in C_1—the *Laminosa* Dolomite. The following horizons are conspicuously fossiliferous: βZ (base of the Main Limestone); γC_1; $C_2 + S_1$; and the base of D.

"In the *Laminosa* Dolomite and *Caninia* Oolite, capped by a thin *Modiola* phase, we have evidence of a Mid-Avonian shallowing which culminated in a brief episode of extremely shallow water. The *Modiola* phase at the base of C_2 which is only exposed at one locality, near Miskin, has a thickness of only 16 feet; but it exhibits a considerable variety of rock-types, namely: calcite-mudstone, limestone-breccia, oolite, contemporaneous dolomite, etc.

"S_2 exhibits a typical development of the *Seminula* Oolite, succeeded by the usual *Modiola* phase with "china-stone" limestones and pisolitic and pseudo-concretionary limestones.

"The *Dibunophyllum* beds are only known west of the River Ely. They have been proved as far east as Bryn-saddler, on the western side of the Ely Valley, by fossils found in the spoil-heap of the old Trecastle Hæmatite Mine; but they are greatly reduced in thickness in that locality, and in all probability they are soon cut out by the continued overstep of the Millstone Grit in an easterly direction. The limestones of this zone attain a thickness of 600 feet, calculated from dip and outcrop, in the west, around Ruthin; and the presence of D_2 beds in that locality is probable. But no evidence of a D_2 fauna has yet been found.

"The junction of Carboniferous Limestone and Millstone Grit is not exposed at the surface. In the workings of the Llanharry Mine, however, where hæmatite is won from the top of the Carboniferous Limestone, the massive limestones of that formation are seen to be succeeded abruptly, but with apparent conformity, by the black shale of the Millstone Grit. There the limestones, which have suffered widespread subsequent dolomitization, contain *Dibunophyllum* and *Cyathophyllum murchisoni*. A thin band of quartzite is developed in the base of the Millstone Grit shale. Nodules of pyritic limestone from the basal shale have yielded a small *Lingula* and an indeterminable coiled cephalopod.

" In this area the ground east of the River Ely, between Miskin and Groesfaen, affords the best opportunities for field-traverses. It is thickly dotted with quarries, which provide sections in every zone except D. Some valuable supplementary sections are afforded by quarries which lie immediately west of the river, due south of Pont-y-clun. Further west, the S_2 beds are particularly well shown in quarries at Llanharry. while the outcrop near Ruthin affords the only good exposures of the D_1 beds.

" 2.—NORTH-EASTERN AREA.

" In a north-easterly direction along the margin of the Coalfield the Carboniferous Limestone undergoes remarkable changes, which may be summarised as follows :—

" (1) *Loss by Overstep.* Continued overstep by the Millstone Grit cuts out the main *Seminula* Zone (S_2), and thus removes a further 600 feet of strata from the top of the Main Limestone.

" (2) *Attenuation of the surviving zones.* The collective thickness of the remaining zones—K (Lower Limestone Shales) and Z to S_1 (Main Limestone)—diminishes from about 1,500 feet to 800 feet. The attenuation is shared, in varying degrees, by all the zones.

" (3) *Increase of dolomites and Modiola phase deposits.* Contemporaneous dolomitization rapidly involves almost the whole of the surviving zones of the Main Limestone. The upper beds, of C—S age, gradually assume the character of a *Modiola* phase of great thickness, composed essentially of dolomite-mudstones with subordinate calcitic beds.

"(4) *Faunal change.* As a concomitant of (3), the standard fauna becomes largely obliterated in the Main Limestone. Only two zonal horizons, the base of Z and the base of S_2 can be recognized east of the Taff valley.

"The vertical sections of fig. 4 illustrate three stages in this transformation of the sequence, and these may be briefly considered.

"(i.) *Taff Valley (Taffs Well and Tongwynlais.)*

"In the gorge of the Taff a nearly complete section is afforded by quarries and railway-cuttings. The sequence may be tabulated as follows :—

			Feet.
Main Limestone	S_2	Oolite and oolitic limestone	380
	C_2 ⊣ S_1	Dolomites, with *Cyrtina carbonaria* at the base	130
	C_1 Z	Dolomites, with two bands of oolite in C. Chert at the base . . at least	1250
Lower Limestone Shales	K	Shales and limestones	100
		Crinoidal Limestone and oolite ..	100
		Limestones and shales, including a *Modiola* phase with limestone of alpha type (*Bryozoa* Bed)	? 60
		Total at least	2020

"The overstep of the Millstone Grit has here progressed little below the top of S_2 while attenuation of the zones as compared with the more westerly outcrops is inappreciable.* But contemporaneous dolomitization has already involved almost the whole of the Main Limestone up to, and including, the basal beds of S_2, leaving unaffected only the bulk of S_2, and two bands of oolite in C. As regards the oolite bands in C, which are recognizable in Garth Wood, on the western side of the valley, the lower band possibly represents the *Caninia* Oolite, while the upper band can be referred to C_2 with certainty. Both of these bands, and the *Seminula*

"* The thickness of main Limestone given in the vertical section of fig. 4 and the above table is probably underestimated. It represents a minimum value calculated from sections in the gorge, where disturbance of the strata militates against accuracy. Recent observations made by the writer [T.F.S.] in Garth Wood, especially an examination of the old workings of the Garth Iron Mine, indicate a great expansion of C_2 and S_1, and a probable total thickness of 2,000 feet for the Main Limestone alone."

Oolite also, have suffered widespread subsequent dolomitization.

"The Lower Limestone Shales of the Taff valley are abundantly fossiliferous, yielding *Cleistopora* cf. *geometrica* and a profusion of brachiopods. In the Main Limestone, however, we find only (*a*) the typical fauna of S_2 (*Semihula ficoides*, *Cyrtina carbonaria*, *Lithostrotion martini*, etc.), (*b*) a fossiliferous band (in the dolomite of Ty-nant Quarry); yielding *Zaphrentis konincki*, *Orthis resupinata*, '*Spiriferina*' cf. *laminosa*, etc., which is referable to Z_2 or γ, and (*c*) traces of the faunas of Z_1 and C_2

"(ii.) *Cefn On.*

"It has been seen that the S_2 beds in the Taff valley comprise some 130 feet of dolomite succeeded by nearly 400 feet of oolitic limestones. Tracing the out-crop in a north-easterly direction from the Taff, rapid overstep by the Millstone Grit soon cuts out the S_2 oolites, leaving only the basal dolomites with a rare band of oolite, to represent this zone.

"The sequence in the Main Limestone can be made out in detail at Thornhill and on Cefn On, about midway between the Taff and the Rhymney. A traverse across Cefn On (fig. 9) gives the following sequence :—

			Feet.
Main Limestone	S_2 (basal beds of)	Dolomites with a band of oolite ..	70
	$C_2 + S_1$ (?) C_1 (part of)	*Modiola* phase	50
		Crystalline dolomite and dolomitic limestone	100
		Modiola phase	250
	C_1 Z	Dolomites, with chert near the base..	450
Lower Limestone Shales	K	(Little exposed) .. .. .. say	170
		Total	1090

"On Cefn On, therefore, the total thickness of the zones K—S_1 is reduced to about 1,000 feet, as compared with at least 1,500 feet in the Taff valley. The Main Limestone is almost wholly dolomitic—some calcitic beds occur in the *Modiola* phases—and even the limestone-division of the

Lower Limestone Shales consists largely of contemporaneous dolomite.

"In this poorly fosiliferous development of the main Limestone between the Taff and the Rhymney, the basal dolomites of S_2, characterized especially by *Cyrtina carbonaria*, form a valuable horizon. Being more resistant to denuding agencies than either the underlying *Modiola* phase or the overlying Millstone Grit Shales, they give rise to a pronounced feature along much of their outcrop.

"(iii.) *Machen and Risca.*

"In the Rhymney valley at Machen, the basal beds of S_2 are cut out by a sharp transgression on the part of the Millstone Grit. The remaining sequence is found to have suffered some further attenuation, reducing the total thickness to 800 feet, made up as follows :—

			Feet.
Main Limestone	C_2+S_1, C_1, Z	.. .. ..	675
Lower Limestone Shales	K	.. .. ..	125
		Total	800

". . . . This brings us to the north-eastern extremity of the area under consideration. We may note, however, that the Carboniferous Limestone continues to be reduced by the overstep of the Millstone Grit in a northerly direction, along the eastern edge of the coalfield. The overstep continues until, at the north-eastern corner of the coalfield, near Abergavenny, little more than the Lower Limestone Shales remains between the Millstone Grit and the Old Red Sandstone. This is the limit of the overstep. In a westerly direction along the "northcrop" of the coalfield the base of the Millstone Grit retrogresses rapidly, with the result that all the zones of the Main Limestone, up to S_2 reappear in a distance of less than six miles.*

* E. E. L. Dixon, *Quart. Journ. Geol. Soc.*, vol. lxxiii, 1918, p. 161.

"3. SOUTH-EASTERN AREA.

"The Lower Avonian is finely displayed in the southern limb of the Cardiff-Cowbridge anticline and the Cadoxton-Barry group of inliers. There the Z—C_1 beds become increasingly fossiliferous, *pari passu* with a loss of dolomitic character, in a southerly or south-westerly direction.

"The zonal range of the strata exposed along the southern side of the anticline, between Cowbridge and Wenvoe, has not yet been completely determined, but it seems improbable that any large part of the Upper Avonian emerges from beneath the Mesozoic deposits which form the southern boundary of the outcrop. At the eastern end of this outcrop the Keuper of the Wenvoe valley isolates the large inlier of Cwrt-yr-ala. In that inlier the highest limestones are of C_1 or lower C_2 age. It may be remarked that the Triassic deposits which isolate this mass and the small inlier at Saintwell, to the north, conceal some pre-Triassic faults of large displacement.

"The Sweldon Quarry, in the Saintwell inlier, exposes highly fossiliferous, cherty limestones of βZ horizon. There, as also throughout the Cwrt-yr-ala inlier, the higher beds of Z are nearly-barren dolomites. But the γC_1 beds, exposed in the Alps Quarry, the Twyn-yr-odin Quarries, and elsewhere, are non-dolomitic and richly fossiliferous.

"In the Cadoxton inlier, the Z beds are still strongly dolomitic, and the *Laminosa* Dolomite overlies the usual non-dolomitic and highly fossiliferous beds of γC_1. At Barry, however, Z has lost all except traces of contemporaneous dolomitization, and even the upper beds of C_1 are but slightly dolomitic.

"The three peninsulas at Barry give a magnificent display of the Lower Avonian in cliffs and fore-shore sections, with triplicate exposures of a great part of the succession. The exposed sequence begins in the upper beds of the Lower Limestone Shales, and terminates a little below the top of C_1. The intervening zones are completely exposed, with some repetition of beds in γ by thrust-faulting. The total thickness of strata in the Main Limestone (β to C) can fall little short of 1,400 feet.

"Outstanding features of the Lower Avonian succession at Barry are the continuously fossiliferous nature of the sequence, the great thickness and extraordinarlly fossiliferous character of γ and C_1, and the crinoidal "Petit Granit" facies of C_1. In these respects the development resembles but surpasses that of Burrington Combe in the Mendip Hills. Chert is stongly developed in Z_1 and at the base of γ.

"The small inliers at Lavernock expose highly fossiliferous strata of γC_1 age.

"4. Western Area.

"Little is yet known concerning the distribution of the zones in the outcrops around the mouth of the River Ogmore. Two outstanding features call for notice —:

"(1) The cliffs at Dunraven, and those between Southerndown and Sutton, give sections of *Caninia* beds beneath an unconformable cover of Lower Lias. The facies is fossiliferous and non-dolomitic, both in C_1 and in Lower C_2. At the Witches' Point the cherty beds of γ are well exposed, and at the Black Rocks the lower beds of C_2 are finely displayed.

"(2) In the extreme west, in quarries at South Cornelau and in the cliffs at Porthcawl, are fine exposures of massive limestones of D_1 age. These beds show a great development of the rock-type termed *pseudo breccia* by the late Mr. R. H. Tiddeman.

"The chert beds at the base of the Millstone Grit, exposed in the Ogmore Valley above Bridgend, are probably of D_3 age, like the radiolarian cherts of Gower, with which they are compared by the Geological Survey.

"It will be clear from the foregoing pages that the outcrops of the Carboniferous Limestone in the neighbourhood of

Cardiff illustrate important changes which take place as the strata are traced from south and west to north and east.

" As regards the varying character of the strata, the principal phenomenon is the lateral change of the Mid-Avonian deposits (C and CS) in a north-easterly direction. This amounts, essentially, to a change from a sequence of "standard" crinoidal and oolitic limestones, interrupted only by a very thin lagoon phase, into a succession of (*a*) contemporaneous dolomites with impoverished standard fauna, and (*b*) lagoon phase deposits chiefly dolomitic in character. The change points to a north-eastward shallowing of this area of the Mid-Avonian sea ; it is accompanied by an attenuation of all the zones which indicates a less rapid subsidence of the sea-floor in the north-east than in the south-west of the area.

" These phenomena compare with the well-known regional attenuation of the Carboniferous Limestone from south to north in the South-Western Province, and the prevailing transition from a deeper-water to a shallower-water facies in the same direction, towards the Avonian shore-line. But in the neighbourhood of Cardiff the changes occur with unusual rapidity ; the significant direction is north-easterly rather than northerly ; and it is evident that we are dealing with local phenomena of special significance. The problem involves a consideration of the evidence supplied by neighbouring areas to the east, which cannot be entered upon here.

" In the unconformity between the Carboniferous Limestone and the Millstone Grit, which is attended by overstep and the gradual cutting-out of successive lower zones of the Avonian in a north-easterly direction along the edge of the coalfield, we have evidence of an episode of intra-Carboniferous earth-movement, emergence, and denudation. The overstep increases to an observed maximum in the north-eastern corner of the South Wales coalfield, near Abergavenny. There, as recorded by Mr. E. E. L. Dixon, the unconformable junction of Carboniferous Limestones and Millstone Grit shows evidence of subaerial erosion of the

Limestone. In a westerly direction along both the "south crop" and the "north crop" of the coalfield, higher and higher zones of the Avonian appear beneath the Millstone Grit as the effect of the earth-movement diminishes." (Quoted from T. F. Sibly *op. cit.*)

An important feature in the Carboniferous Limestone of the Cardiff District is the local replacement of limestone by hæmatitic iron-ores similar to the well-known occurrences in the Furness District. The replacement was caused by water percolating downward from a former extension of the Triassic cover. As one would naturally expect, particular beds in the limestone were particularly susceptible to lalteration by reason of their texture, or some other factor. But the change has taken place along any fissure, bedding-plane, joint fault, or other opening that served to let in the waters. Thus the ore-bodies, while often guided by the bedding, are actually highly irregular and they usually fail in depth. Iron-ore is now worked at Llanharry and was formerly worked in Garth Wood, near Taffs Well, while trials have been made at a number of places on either side of the Taff Valley. In addition to iron-ore, barytes and galena occur in irregular veins and patches in the Limestone, but hardly in workable quantity.

References. See list on p. 1; also.
VAUGHAN, A.—"The Carboniferous Limestone Series (Avonian) of the Avon Gorge." *Proc. Bristol Nat. Soc.*, 4th Series, vol. i., part ii. (1906), p. 74.
——*Rep. Brit. Assoc.* 1909 (Avonian Zones, etc.), p. 187.
——"Correlation of Dinantian and Avonian." *Q.J.G.S.*, vol. lxxi. (1915-16), p. 1.
REYNOLDS, S. H., and VAUGHAN, A.—"The Avonian of Burrington Combe (Somerset)" *ibid.*, vol. lxvii. (1911), p. 342.
DIXON, E. E. L., and VAUGHAN, A.—"The Carboniferous Succession in Gower (Glamorganshire)," *ibid.*, p. 477.
DIXEY, F., and SIBLY, T. F.—"The Carboniferous Limestone Series on the South-Eastern Margin of the South Wales Coalfield," *ibid.*, vol. lxxiii. 1918 p. 111.
SIBLY, T. F.—"Hæmatites of the Forest of Dean and South Wales." Mem. Geol. Survey, 1919.

(*ii.*) *Upper Carboniferous* (*a*) *Millstone Grit Series.*

The Millstone Grit Series is only seen in the Coalfield Rim where its outcrop forms a rather narrow band. The succession is most complete in the North Crop, where it is as follows :—

Farewell Rock ..	Quartz-conglomerate or massive sandstone passing downwards into alternations of sandstone and shale	150—200 feet
Shales	Dark shales with thin seams and nodules of clay ironstone also occasional grit bands	100—150 ,,
Pebbly Grit ..	Quartz-conglomerate with lenticular partings of dark shale	80—100 ,,

The rocks form a moorland dip-slope deeply trenched by rivers that flow down to Glyn Neath and elsewhere, the sandstone bands causing a number of waterfalls, especially at localities where there is displacement by faulting. In places the moorlands are studded with swallow holes and the drainage is carried underground. through solution channels in the underlying limestone.

Along the South Crop the outcrop is much narrower, partly owing to the high dip and partly owing to the local overlapping of the pebbly basement beds. Under these circumstances the outcrop lies along a depression, with a small feature marking the harder Farewell Rock (fig. 7).

The series yields a marine fauna, goniatites, lamellibranchs, etc., along with drifted plant remains. Vegetable material is sometimes abundant enough to form thin coal smuts or even occasionally thin coal-seams. As regards the marine fossils, most of the fossil localities are in the North Crop, where the series is best exposed and where decalcified limestone bands are developed in the shales.

The basal quartz-conglomerates are extensively quarried in the Glyn Neath district and elsewhere for the preparation of silica-bricks. Along the South Crop the pure white quartzite and conglomerates of the Farewell Rock have also been worked for a similar purpose.

(*b*). *The Coal Measures.*

The Coal Measures of South Wales are divisible into four divisions as follows :—

Upper or Llantwit and supra-Llantwit Measures	Shales, sandstones, with some ironstones and coal-seams, with the Mynyddislwyn Seam at the base.
Pennant Series	Felspathic micaceous flags and sandstones often conglomeratic, with subordinate shales, and, west of the Taff, some thin and sulphury coal-seams.
Lower or Steam Coal Measures	Shales and fire-clays, with bands of white quartzite, felspathic and conglomeratic sandstones, and the principal ironstones and coal-seams.

The characteristic physical features made by the crops of the different divisions have already received notice as also the remarkable eastward thinning apparent in every division of the Carboniferous System.

The Lower Measures.

The Lower Measures are, of course, the most important economically, as they contain nearly all the workable coal-seams, as well as the ironstones that were worked in former years. Like the productive Coal Measures of other Coal-fields, the strata are a series of shallow-water estuarine deposits, remarkable for their great lithological variety—shales of various types, fireclays, pure white quartzites, felspathic and conglomeratic sandstones, coal-seams, etc., succeeding one another in rapid alternations. A certain sequence is often observable, sandstone being followed by shale, shale by clay, clay by coal ; above the coal may come more shales, then sandstones, and once more the cycle is repeated. The sandstones evidently represent shallower water conditions, with maximum current action, while the coal-seam marks the stage of maximum subsidence, or minimum current action. At their maxima the currents were often powerful enough to cause erosion ; accordingly conglomeratic sandstones may rest directly upon a coal-seam and cut out the shales that might be expected to occur above the seam, or they even cut out the seam itself. The roofs of certain seams are particularly prone to variations. Such conglomerates may contain coal pebbles, which prove that the transformation of vegetable material into coal was already accomplished before the conglomerates were formed.

The number of workable coal-seams varies in different districts. As a rule there are about 10 important seams, totalling some 45 feet in average thickness. As usual in Coalfields the seams receive different names in different districts, their quality alters laterally, and individual seams disappear by attenuation, so that correlation of the seams throughout the Coalfield is a matter of great difficulty, if not impossibility.*

The westward passage of the seams from bituminous and steam-coals to anthracites has been the subject of frequent discussion, but no complete explanation of the change has yet been put forward. The change usually becomes first apparent in the lower seam. It has, nevertheless, been proved that the change is not connected with increased depth of burial, and it would seem that the present differences in composition are due to original differences in the materials of which the seams are composed. Further application of recent combined palæobotanical and chemical research methods may be expected to furnish the real clue to this problem.

The ironstones are clay-ironstones present as definite bands or "mines" and as concretions in the shales and fireclays. It is a point of historic interest that the original workings in the Coalfield were for the purpose of obtaining iron-ore, which, in those days, was then smelted with wood-charcoal. Only as the timber was exhausted were the coals worked.† Along the crop the mine-ground was sometimes cleared by scouring with a sudden rush of water. Traces of such workings may be seen near Caerphilly. Nowadays the clay-ironstones have ceased to be of commercial importance owing to the importation of purer ores from abroad and the opening up of the more easily mined Jurassic ores of the Midlands.

The palæontology of the South Wales Coal Measures long remained an almost untouched field for investigation. Plant remains, though often scarce throughout considerable thicknesses of strata, occur abundantly at particular horizons. Shell-bands are not so common in South Wales as in many Coalfields, but they do occur along with some of the coal-seams of the Lower Coal Measures, and they suggest a marine origin for the strata.

---

* For table of synonyms see H. K. Jordan, "The South Wales Coalfield Section and Notes." *Proc. South Wales Inst. Eng.*, vol. xxvi. (1908), p. 56.

† A. E. Trueman, "Population changes in the Eastern part of the South Wales Coalfield." *Geographical Journal* (1919), p. 410.

The Pennant Series.

The Pennant Series is a Formation peculiar to the South Wales Coalfield and the adjacent Bristol and Forest of Dean Coalfields. Scenically it may be described as the characteristic formation of South Wales, both on account of its extent, and the manner in which it fominates all other features. It consists of a great thickness of felspathic sandstones, sometimes conglomeratic, often micaceous and flaggy and green to grey in colour. Shales are sometimes intercalated with the sandstones, especially in the central portion of the series. As the formation expands westwards, in accordance with the general rule in South Wales, thin coals develop in this shale group and eventually become of economic importance. The coals first develop near the Taff Valley—the Coed-cae-dyrys and Stinking Veins—but do not attain their maximum until much further west in the Swansea District. These coals are, as a rule, highly sulphurous.

The sandstones are often false-bedded and contain frequent lenticular coal-partings, some of which clearly represent detrital material, showing that somewhere in the neighbourhood coal-seams were already undergoing denudation.

The upper boundary of the Pennant Series is defined by the Mynyddislwyn coal-seam of Monmouthshire, and its lower boundary by the Rhondda No. 2 seam of Glamorganshire, equivalent to the Tillery seam of Monmouthshire. This is not always a satisfactory definition in view of the known difficulty of correlating the seams of different areas. Additional complication is caused by the frequent presence of sandstones of Pennant type, the Llynfi rock of the Swansea District, in the upper portion of the Lower Measures. As examples may be cited, in the South Crop the sandstones on Caerphilly Common, and in the North Crop, those under Craig-y-Llyn.

The Upper or Llantwit and Supra-Llantwit Measures.

Above the Pennant sandstones comes a further series of shales and sandstones, with some ironstones and thin coals, the Llantwit seams. These rocks are only preserved in synclinal basins, such as the Caerphilly Basin (fig 2). In the basins of the eastern half of the Coalfield, some 300 feet of these beds have escaped denudation. Further west, in basins at Bryn Coch, near Neath, and at Gorseinon, in the Swansea District, a much greater thickness of Upper Measures is preserved, including beds higher than any in the Cardiff District. These highest strata of the eastern basins have been termed the Supra-Llantwit Measures, and they represent the stratigraphically highest Palæozoic Rocks in South Wales. They comprise a series of red strata, almost barren of coal-seams.

So far in this paper the terms Lower (Steam Coal) Measures and Upper (Llantwit) Measures have been used in their purely local sense as employed in the South Wales Coalfield. Using the plants as a guide, the correlation with the Coal Measures of other areas seems as follows :—

Local Divisions.	*Palæontological divisions (Kidston and others).*
Upper Coal Measures of South Wales or Llantwit and Supra-Llantwit Measures	Upper Coal Measures or Radstockian.
Pennant Series	Transition Measures or Staffordian.
Lower Coal Measures of South Wales, or Steam Coal Measures	Middle Coal Measures or Westphalian.
? absent in South Wales ..	Lower Coal Measures or Lanarkian.

These above correlations must, however, only be regarded as approximate. For example, it must not be supposed that the base of the Pennant coincides exactly with the base of the Transition (Staffordian) Coal Measures ; in the central part of the Coalfield the Transition Coal Measure flora makes its appearance below the base of the Pennant. The at present unpublished work of Mr. David Davies may be expected to shed much light on this question. Further, the lithological zones appear to transgress the palæo-botanical zones. Thus the Pennant Series transgresses from lower to higher floral horizons as it passes eastwards from Pembrokeshire through Glamorganshire and Monmouthshire to the Forest of Dean ; * in the latter area the " Pennant " rocks are all of

* T. F. Sibly, *Q.J.G.S.*, vol. lxix. (1913), p. 278.

Upper Coal Measure (Radstockian) age according to Newell Arber.*

A puzzling feature is the apparent absence of local representatives of the "True" Lower or Lanarkian Measures of Dr. Kidston. It might be expected that such representatives are to be looked for in the local "Millstone Grit." The palæo-botanical zone of the Glamorganshire Millstone Grit has not been established, but in Pembrokeshire the "Millstone Grit" contains, as far as can be determined, a Middle (Westphalian) Coal Measure flora.† The break below the Millstone Grit of Glamorganshire and Monmouthshire may account for this big gap, and it is possible that there are other breaks yet undetected at higher levels. As already stated, our knowledge of the palæo-botany of the South Wales Coal Measures is remarkably incomplete and there is abundant scope for further research.

As regards the lithology, it seems often taken for granted that the Coal Measures of South Wales are so distinctive that no exact comparison with other Coalfields can be instituted. There is much to be said for such a view ; nevertheless, certain general correspondences are very striking, even though there may be great differences in detail. We may take the sequence in the South Staffordshire Coalfield as a basis for comparison as set out in the following table :—

South Wales.		*South Staffordshire.*	
Supra-Llantwit Series	Red sandstones and marls with purple shales.	Keele Group	Red sandstones and marls with *Spirorbis*-limestone bands. ? *Local unconformity.*
Llantwit Series	Shales and sandstones with thin coals ; Mynyddislwyn seam at the base.		
Pennant Series.	(iii.) Grey sandstones and flags. (ii.) Shales with thin coals. (i.) Grey sandstones and micaceous flags with some conglomerates.	Halesowen Sandstones	(iv.) Grey, brown, yellow sandstones and shales with coaly traces. (iii.) Red, brown, and grey sandstones. (ii.) Blue clay with *Spirorbis* limestones and thin coal. (i.) Sandstones and conglomerates. *Local unconformity.*

* "The Fossil Flora of the Forest of Dean coalfield," *Phil. Trans.* Ser. B., vol. 202 (1912), p. 272.

† R. H. Goode, "The Fossil Flora of the Pembrokeshire coalfield, *Q.J.G.S.*, vol. lxix (1913) p. 275.

	South Wales.	*South Staffordshire.*	
Steam Coal Measures	(iii.) Ronddda No. 2 Seam; Rock veins of Caerphilly district with associated shales and conglomerates.		
	(ii.) Red and purple clays with green grits of Espley type.	Old Hill Marls.	Red purple clays with green grits, sometimes conglomeratic (Espley rock).
	(i.) Shales and fireclays with some grey and white sandstones and conglomerates, also clay ironstones and numerous coal-seams.	Productive Measures	Shales, fireclays, ironstones and coal seams.
True Lower Coal Measures or Lanarkian	?*Absent.*	*Absent.* (unconformity.)	
"Millstone Grit"		*Absent.*	

The plants show that there is a general palæo-botanical correspondence between the major divisions in the two areas, while the lithological sequence is seen to be somewhat similar even as regards some of the sub-divisions. This is not to say that there are no considerable differences in detail. Such differences are only to be expected in view of the greatly diminished thickness of the South Staffordshire Coal Measures. It should, however, be noted that the difference in thickness between Staffordshire and Monmouthshire is no greater than the difference between Monmouthshire and West Glamorganshire.

Towards the end of Carboniferous times the Armorican movement caused great elevation, folding and denudation. Accordingly we do not know whether any Palæozoic strata ever succeeded the Upper Coal Measures, which are themselves only preserved in synclinal basins. The relations of the Mesozoic Rocks to the Palæozoic prove that the structures in the latter rocks were already in pre-Mesozoic time substantially as we now find them.

Structure of the Coalfield.

The Coalfield forms, geologically speaking, a great basin elongated east and west about 16 miles broad and (excluding the detached Pembrokeshire Field) rather over 50 miles long. It is essentially a "visible" Coalfield, only a very small portion of the South Crop near the Ely Valleys is "concealed" by Mesozoic Rocks. The major portion of the outcrop is occupied by the Pennant Series, while the lower productive measures appear towards the north and south margins, the "North and South Crops." The general structure is shown in the section fig. 2 and is seen to be markedly asymmetrical; dips are steep along the South Crop, quite gentle along the North Crop. As a result of the asymmetry the deepest parts of the basin are situated quite near the South Crop. The basin is, moreover, not a simple syncline, but is divided by a strong anticline, the Pontypridd Anticline, which is of great importance in that it serves to keep the seams within workable distance from the surface over a wide area. North and south of this anticline are the synclinals of Gelligaer, and of Caerphilly and Llantwit, which bring on the Upper Coal Measures. These synclinals are usually referred to as the "North" and "South Troughs" respectively. The Pontypridd Anticline does not keep strictly parallel to the Coalfield Rim, but takes a somewhat curving course. Nor is it everywhere equally developed; westwards it passes gradually into an overthrust which runs towards Swansea Bay

Other important structures in the Coalfield are the N.E. or E.N.E. disturbances, of which the best example is the Vale of Neath disturbance, a long narrow belt of compression which in some places shows as folding, in others as faulting. It thus resembles the Pontypridd Anticline, in that it shows the lateral passage of an anticline into an overthrust. Its effect on the river system has already been noted. Other similar disturbances occur west of the Neath disturbance, but not so prominently east thereof, except for one near Pontypool which runs towards the Pontypridd Anticline.

All parts of the Coalfield are traversed by a set of N.W. or N.N.W faults, which are very numerous, but mostly of small throw. Some of these seem to stop short at the Mesozoic Rocks, but others penetrate the newer rocks and appear on the coast. The Taff Gorge is excavated along the line of one of the most powerful of these N.N.W. faults.

The main fault is here accompanied by a number of subsidiary faults. Underground workings prove the local existence of other sets of faults in different areas, sometimes forming a regular plexus, but their relations to other tectonic features has received but little notice.

Locally there exist peculiar rolls in the strata, sharp folds which pass into overfolds and faults, and cause difficulties in working the coals. The strata affected are usually highly slickensided. In any one district such rolls may be very common along a given horizon and yet fail to affect the beds above and below. They are evidently connected with the major folding and do not occur in areas unaffected by the major folds. They seem especially prevalent where hard beds alternate with soft shaley strata, but they have not yet received detailed examination.

Much work yet remains to be done on the tectonics of the Coalfield before we have a clear picture of the relationships between the folding and faulting.

The Mesozoic Unconformity.

It has already been mentioned that consequent on the Armorican uplift and folding, an irregular land surface was carved out, and that subsequent subsidence resulted in deposition of the Mesozoic rocks, which rest with a great unconformity on the denuded edges of the older rocks, and still remain almost horizontal (fig. 1 and fig. 4). But as the Mesozoic subsidence proceeded the succeeding Formations were deposited over a continuously enlarging area. Accordingly we find that each member of the Mesozoic Series overlaps the member below it, and eventually oversteps directly on to the older rocks. This overstep takes place westwards through Glamorganshire. Thus around Cardiff and Barry the base of the Mesozoic is in the Keuper Marls, westwards near Bridgend, Rhætic rocks form the base, and only a short distance away along the Southerndown Coast, both Keuper and Rhætic are missing and Liassic Rocks rest directly upon the Palæozoic. The overstep results in great changes in the lithology of each Formation as it is followed across the county. For as a Formation oversteps on to the older rocks, in other words, as it approaches the old coast-lines, it loses its normal character and assumes a littoral habit. Thus we may see clays passing laterally into lime-stones and these in turn into sandstones and conglomerates. Similar lithological changes take place northward with approach to the high ground of the Coalfield Rim, and the northward changes are even more rapid than the western.

As already mentioned, Mesozoic Rocks are restricted to the Vale of Glamorgan, but the Coalfield Rim was broken down in Triassic times at one locality (what is now the Ely Valley) where Trias rests on Coal Measures. The disposition of the various Formations shows that scarps round the Coalfield Rim were already in existence in Triassic times, for the Triassic Beds are seen to run up the old hollows on the Millstone Grit Shale and Lower Coal Measure outcrops.

4. TRIAS: *a. Keuper Marls Series.*

The lowest Mesozoic rocks in the district are the Keuper Marls; no representatives of the Bunter and Keuper Sandstone of the Midlands are present in South Wales. The full development is as follows:—

Rhætic Beds.	Grey Marls	..	up to 20 feet.
Keuper Marl Series.	Tea-green Marls	..	,, 25 ,,
	Red Marl ..	..	,, 400 ,,
	Littoral Beds	..	,, 60 ,,

Unconformity.

There is a complete upward passage and the various divisions grade one into the other without any sharp line of demarkation.

The Littoral Beds, or Keuper Basement Beds, are a type peculiar to the districts that border the Bristol Channel. They represent a marginal or basal facies of the Red Marls, developed between the Red Marls and the older rocks upon which they rest. As a result of the Mesozoic overlap the Littoral Beds develop at different horizons in different districts, so that westwards, where the Red Marls have been overlapped, the Tea-green and finally the Grey Marls are represented by Littoral deposits.

They comprise a set of breccias, conglomerates and limestones with interbedded sandy or calcareous red marls. Their exact character varies according to the beds on which they rest. Thus where the Keuper Marls rest only upon red marls of the Old Red Sandstone, as at Cardiff and Newport, the basement conglomerates and breccias are not developed, but where the Keuper Beds were laid down over, or in proximity to areas of Old Red Sandstone, Brownstone and Quartz-conglomerates, then the Keuper Basement Beds are mainly made up of re-arranged Old Red Sandstone material. The "Radyr Stone" formerly much used for rougher building work in Cardiff, furnishes an example of this type. Where the beds rest on Carboniferous Limestone as at Sully, Barry, and elsewhere, the Basement Beds consist of Limestone-conglomerates and breccias (the "Dolomitic" conglomerate of the Bristol District). The breccias are sometimes exceedingly coarse. They do not necessarily rest upon an even surface of the older rocks; on the contrary, they may frequently be seen banked up against steep pre-Triassic cliffs, as at Nell's Point, Barry Island, and elsewhere.

The breccias may be only a foot or two thick, or may be as much as 60 feet. Above often come grey or reddish fine-grained limestones without any trace of organic structure and evidently of truly detrital origin, derived from the waste of Carboniferous Limestone. Such rocks are often difficult to distinguish from the true Carboniferous Limestone, especially as they may contain fossils washed out of the Carboniferous Limestone. Surfaces are sometimes ripple-marked. Intercalated with the limestones are beds of normal red marl and of hard calcareous marl, showing all transitions to limestone. Other conglomeratic bands may come in actually among the limestones and hard marls. The coast from St. Mary's Well Bay past Sully to the Bendrick Rock gives a fine section of the Littoral Beds seen to a total thickness of 50 or 60 feet, and mostly made up of limestones and hard marls with a smaller amount of conglomerate. At Sully Island and Bendrick Rock the unconformity is well seen, likewise on Barry Island, on either side of Whitmore Bay.

The Red Marls show the usual altenation of soft red marls with harder bands, and as usual with almost any red series, the colour may change to green along definite bands, or in irregular spots and lenticles. Bands of gypsum develop in the higher part of the series just as in the Midlands. The gypsiferous beds and the overlying green and grey marls are well seen in Penarth Cliffs. Borings made at various places between Cardiff Docks and Penarth prove the total thickness to be rather over 400 feet, or much less than in the Midlands. With approach to the old shore lines the marls pass laterally in the littoral facies described above.

The Red Marls pass upwards by alternation into the Tea-green Marls and these into Grey Marls, but there is no sharp line of division between Red and Green Marls, still less between Green and Grey Marls. Both Green and Grey Marls show an alternation of soft marls and hard bands similar to that in the Red Marls, and gypsum bands are again developed in the Tea-green Marls. Both Green and Grey Marls pass westwards into a littoral facies.

At a few localities a calcareous band up to three feet in thickness containing *Ostrea-bristovi* is found above the Grey Marls and below the Rhætic Beds. At the coast this band is only seen in the wooded slope at St. Mary's Well Bay. It also occurs at a few inland localities nearer Barry. It is not found elsewhere along the coast, as it is cut out by the plane of erosion that brings on Rhætic Beds. The importance of this band is that with its marine fossils it marks the first incursion of sea-waters into the Keuper Lake. The band was linked by Mr. Richardson* to the Grey Marls to the upper portion of which he applied the term "Sully Beds,' and classified these Sully Beds with the Rhætic Series, not with the Keuper. The term is not happily chosen, as the beds in question are found only in St. Mary's Well Bay, not at Sully itself, which is on the outcrop of Keuper littoral beds.

* "The Rhætic . . . . of Glamorganshire." *Q.J.G.S.*, vol. lxi. (1905), p. 385.

Further the base of the Sully Beds is not defined, and there is a perfect transition from them to the Tea-green Marls below. Accordingly the classification adopted by the officers of the Geological Survey is used in this paper. It may be stated that an exhaustive examination of the Trias Rhætic junction along the coast-sections has been recently made by Mr. F. F. Miskin. The results are at present unpublished, but they strongly support the grouping of the 'Sully Beds' with the Keuper Series. Above the *Ostrea bristovi* Bed where present, otherwise above the Grey Marls, comes a plane of erosion which marks the base of the Rhætic Beds.

As elsewhere in the country the Keuper Marl Series is almost unfossiliferous, since the waters of the Keuper Sea were evidently too saline to support much life. Apart from the *Ostrea bristovi* of the topmost band, a few fish teeth have been obtained from the Grey Marls five feet below the base of the Rhætic, at Goldcliff, near Newport, also near Sully and Dinas Powis. Small reptilian teeth have been obtained in the marls six feet below the Rhætic at Lavernock, and reptilian footprints were discovered on one of the basement breccias.

b. Rhœtic Series.

The Rhætic Beds fall naturally into two divisions—the Black Shales, or *Avicula contorta* Beds below, rather over 20 feet thick, and the White Lias above, about 11 feet thick. The localities around Penarth and Lavernock are classical for the study of these beds, formerly termed the "Penarth Beds." Nowadays the best exposures are found in the cliff-sections, between Penarth and Lavernock, and at Lavernock Point. The sections at Penarth Head, and the inland exposures are not so clear. Other exposures occur at Barry Dock Station.

Below the Black Shales, separating them from the Keuper, is the plane of erosion already mentioned, on which rests the basal bed of the Black Shales. This is a "bone-bed" particularly interesting both by reason of its lithology and its fossil contents. It is usually not more than an inch or so thick and is not absolutely persistent, but occurs rather in pockets eroded out of the underlying marls. It partakes sometimes the characters of a conglomerate or breccia with rounded quartz-pebbles occasionally half an inch or more in diameter, usually less, also rounded and angular marlstone pebbles derived from the underlying Keuper. Frequently it consists largely or almost entirely of reptilian coprolites, made up largely of quartz-grains and small pebbles, together with numerous fish teeth and scales. Normal black shale often wraps round these coprolites.

The succeeding Black Shales contain several thin limestone bands, some of which are sandy and ferruginous with abundant fish teeth and scales, "bone beds"; others are purer, others again shell marls. The shales themselves and some of the limestone bands are abundantly fossiliferous, yielding *Avicula contorta*, *Pecten valoniensis*, *Protocardium phillipianum*, and other characteristic Rhætic fossils. The White Lias is made up of white or pale coloured marls with intercalated bands of limestone and grit. Some of the limestones are very compact, smooth and fine-grained, others are largely composed of shells, including *Pecten*, *Protocardium*, *Ostrea*, *Lima valoniensis*, *Modiola minima*, *Plicatula intusstriata*, and other characteristic forms. The strata bear every evidence of shallow water origin; the grit is well ripple-marked, sometimes in three directions, and the higher beds show sun-cracks. The cracks are now sealed with calcium carbonate or with a calcareous marl. As often the case in shallow water deposits, the fossils are frequently represented by small forms crowded together along particular bands, and the state of preservation is poor.

The Rhætic Beds preserve their normal aspect, near Newport and round Penarth and Barry, but assume a littoral facies around Cowbridge and Bridgend. The shales and marls pass laterally into massive oolitic limestones and finally into massive sandstones.

5. Lower Lias.

The Rhætic Beds are succeeded conformably by those of the Lower Lias, which are well exposed both along the coast and in inland quarries. In the immediate neighbourhood of Cardiff they occur in several outliers, the chief of which are Penarth, Lavernock, Leckwith, and St. Fagans (see map, fig. 9). West of Barry is a much larger outcrop, which stretches for nearly 20 miles westwards to beyond Southerndown and Bridgend, and extends over the greater part of the Vale of Glamorgan.

In the outliers of Penarth, Leckwith, and Lavernock, and also in the area just west of Barry, the Lias shows a normal facies of limestones and shales. But when followed further westwards, the beds pass laterally into a littoral facies of massive limestones and basal breccias. It is convenient to consider first the succession in those areas where the facies is normal.

The general sequence was first worked out in the cliffs at Penarth, where the following succession is clearly shown:

E. Limestone and Shales about 50 feet.
D. Shales ("Lavernock Shales") about 40 feet.
C. Limestone and Shales about 45 feet.
B. White Lias about 9 feet.
A. Rhætic and Keuper.

Wright* and H. B. Woodward† examined these beds and suggested that the lower limestones and shales (C in the above table) represented the *planorbis* zone, while the divisions D and E, the Lavernock Shales and the overlying limestones, were referred to the *bucklandi* and *turneri* zones. Mr. L. Richardson made some more exact observations and was able to determine the extent of some of the zones, but beyond introducing a more modern terminology, his section, which, of course, deals mainly with the Rhætic Beds, does not

* "On the Zone of *Avicula contorta* and the Lower Lias of the South of England," *Q.J.G.S.*, vol. xvi. (1860), p. 374. Also in "Lias Ammonites," 'Palæontographical Society (1881), p. 271.

† "Notes on the Rhætic Beds and Lias of Glamorganshire," *Proc. Geol. Assoc.*, vol. x. (1888), p. 529; "The Jurassic Rocks of Britain," vol. iii. (1893), *Mem. Geol. Survey*,.

give much additional information concerning the Lias succession.*

The zonal groupings used by these workers are shown in the following table, along with the classification proposed by Dr. A. E. Trueman.†

Lithology.	Wright.	H. B. Woodward.	L. Richardson	A. E. Trueman.
E. Limestone and shale	Zones of *A. turneri and*	Zones of *A. semicostatus*	*? rotiformis*	*angulata.*
D. Lavernock Shales	*A. bucklandi*	*A. turneri and A. angulatus.*	*marmorea*	
C. Limestones and shales	Zone of *A. planorbis.*	Zone of *A. planorbis.*	*megastomatos.*	*Wœhneroceras johnstoni.*
C. Limestones and shales			*planorbis.*	*planorbis.*
C. Limestones and shales			*Ostrea.*	*Ostrea.*

"Concerning the lower beds, which contain no ammonites, but are rich in *Ostrea liassica* Strickland, there is little difficulty. Above these occur a series of limestones and shales, which are generally referred to the *planorbis* zone, and which contain ammonites of the *planorbis* and *johnstoni* types. It should be noticed, however, that smooth *planorbis*-like forms are confined to the lower part of these beds, while the ribbed *johnstoni*-like ammonites occur solely in the upper part, so that two well-marked sub-divisions can easily be distinguished.

". . . . The overlying Lavernock Shales are richly fossiliferous wherever exposed, the lowest beds containing several species of *Wœhneroceras*, while the main mass of shale yields abundant examples of *Schlotheimia angulata* and allied species. . . . . No positive evidence of any zones higher than the *angulata* zone has been obtained.

"The rocks may be most completely studied at Lavernock, where they are well shown in cliff sections, and where many

* "The Rhætic and contiguous Deposits of Glamorganshire," *Quart. Jour. Geol. Soc.*, vol. lxi., (1905), p. 385.

† "The Liassic Rocks of the Cardiff District" *Proc. Geol. Assoc.*, vol. xxxi. (1920), p. 95.

of the horizons may be examined on the beach, whereas at Penarth, the cliffs are not easily accessible.

"The Lavernock outlier consists of an oval patch, two and a half miles in length, in which the rocks are folded into a gentle syncline along an axis trending roughly north and south, which is shown on the south in cliff-sections (fig. 10). At either end of these cliffs the junction with the Rhætic beds is seen, and proceeding inwards towards the centre of the syncline successively younger beds appear in the cliffs and outcrop on the beach, where most of the limestone bands project as reefs." Quoted from A. E. Trueman *op. cit.*

The general succession is as follows :—

		Ft.	in.	Fossils.
angulata beds	Shales with many layers of orange-stained nodules about	50	0	*Schlotheimia* spp. *thalassica*, *Ostrea irregularis*, *Cardinia ovalis*.
	Dark shales with occasional nodules (Lavernock Shales)	40	0	*Schlotheimia* spp., *Cardinia ovalis*, *Ostrea irregularis*.
Wæhneroceras beds	Shales with several fairly massive limestone bands	6	0	*Wæhneroceras* spp., *Caloceras* sp.
johnstoni beds	Alternating shales and nodular limestones	19	0	*Caloceras* spp., *Lima gigantea*.
planorbis beds	Dark-blue shales with thin limestones	4	7	*Psiloceras* cf. *sampsoni*
Ostrea beds	Alternating limestones & shales about	21	0	*Ostrea liassica*, *Modiola* sp., *Saurian bones*.
	Paper Shales	1	3	
White Lias	Light-grey marls and shales	9	2	*Plicatula intus-striata*.

The lowest beds, the White Lias and Ostrea beds, contain no ammonites, the earliest ammonites appearing in the *planorbis* beds. These ammonites are for the most part smooth but badly preserved. The ammonites of the *johnstoni* beds are ribbed.

The succeeding divisions contain angulate ammonites, the lower group with *Wæhneroceras*, a type that does not acquire the channelled venter characteristic of *Schlotheimia angulata*.

Other mollusca are abundant in these beds ; *Lima gigantea* and allied species, *Modiola*, *Pleuromya*, etc., are common, also oysters of various types. In the lowest beds the oysters are relatively flat and of the *Ostrea liassica* type, but in higher beds the height of the valves increases, as in *O. irregularis*, which in the upper part of the *angulata* zone give place to forms of *Gyryphea*.

Saurian remains are also found, and many crinoid stems and echinoid fragments.

"The lithological characters of the various sub-divisions of these rocks give rise to the physical features of the coast. For instance, small headlands are formed by the relatively hard limestones of the *Ostrea* Beds, and between these headlands the Lavernock Shales, brought down to the beach by the syncline, have been worn back more rapidly and give rise to a shallow bay. Nevertheless the upper part of the cliff in the centre of the syncline, where limestones again predominate, is once more almost vertical. At the extreme ends of the section where the Rhætic beds form the cliffs and foreshore, the cliffs are again worn back, producing the cove at Lavernock, in the east, and St. Mary's Well Bay in the west. In the latter bay the basement beds of the Keuper are brought up against the Rhætic and Liassic rocks by a rather complicated fault.

"Some of the reefs formed by the limestones on the beach are wide, and may be followed at low tide for nearly a quarter of a mile towards the sea. As fossils are very abundant in many of these reefs, specimens may be collected in place without much difficulty. The Lavernock Shales and higher beds, however, do not outcrop on the beach, and consequently cannot be examined so readily, so that the sequence in these beds has not been determined in so much detail.

"The limestone beds are of two main types, which give rise to reefs of very different appearance. A few limestones are evenly-bedded, and stand out on the beach as flat pavements, cut into slabs by several sets of joint planes. Many limestones, particularly in the higher beds, are nodular, and when uncovered the tops are very irregular. Dr. Lang drew attention to the occurrence of these two types of limestone at Lyme Regis, and suggested that both are due to the secondary segregation of the calcium carbonate.*

"Apart from the excellent exposures on the beach at Lavernock, there are many inland exposures in the same outlier, for the lower beds are worked in several quarries by the South Wales Portland Cement Company, while in a cutting a few yards north-east of Lavernock Station the folded Lavernock Shales are again exposed. The distribution of the various zones may be traced inland by the features to which they give rise ; the *Ostrea* Beds form a low, but well-defined escarpment, while the Lavernock Shales give rise to a belt of heavier ground, the upper limestone series again forming higher ground, both near the coast and to the north, at Down's Wood. . . . .

"In the Penarth outlier the same series of beds is present, showing a complete transition from Keuper Marl, through Rhætic, to the limestones above the Lavernock Shales. The section may be examined in the cliffs, just in front of the church, where it is moderately easy to climb. The rocks are faulted and it is therefore quite possible that somewhat higher beds are present than are exposed at Lavernock, . . . but no ammonites of "*bucklandi*" pattern have been discovered.

"The lower zones were formerly worked extensively at the western end of the outlier, but there are now no sections there worth visiting. The sequence at Leckwith is apparently quite similar to that at Lavenock, but except for quarry sections in the lower zones, there are few good exposures." Quoted from A. E. Trueman, *op. cit. supra.*

The large area of Liassic rocks west of Barry illustrates the lateral passage from a normal to a littoral facies. This area is now being investigated zonally by Dr. Trueman. At Porthkerry, just west of Barry, the whole sequence from the *Ostrea* Beds upwards may be studied. Further west, shales which resemble the Lavernock Shales are seen at the base of the cliffs, as for example east of Llantwit Major.

* W. D. Lang, "The Geology of the Charmouth Cliffs, etc.," *Proc. Geol. Assoc.*, vol. xxv. (1914), p. 297.

Above these shales come massive fine-grained limestones which build up the greater part of the cliffs and are at least 200 feet thick. They are often comparatively unfossiliferous but . . . . "they are seen to include several zones higher than those represented near Cardiff. Many of the ammonites are probably new, but they are generally Arietids of several types, formerly called *Ammonites bucklandi*, while in the upper beds some curious evolute Arnioceratеs are sometimes abundant. Beds of lamellibranchs, chiefly *Gryphea*, which is often beekitised, are common.

"Further west, the base of the Lias is again seen in the cliffs near Southerndown and Dunraven Castle, resting unconformably on Carboniferous Limestone.* These deposits were laid down in the neighbourhood of islands of Carboniferous Limestone, and consequently these basement beds of the Lias are very different from the corresponding rocks further east. With the complete submergence of the Carboniferous Limestone islands, however, the Liassic deposits naturally become more normal.

"The succession seen at Southerndown is typical and, perhaps, the best known; two divisions in the littoral deposits are recognised there. Resting on the Carboniferous Limestone are some forty feet of white limestone, often conglomeratic, known as the Sutton Beds, which probably constitute a littoral modification of the *Ostrea* Beds and the lower part of the *planorbis* zone, although the basement beds in different localities may not be of precisely the same age. Above the Sutton Beds occurs a variable thickness of darker conglomeratic limestones, often containing chert, known as the Southerndown Beds, which appear to represent some part of the *angulata* zone. Above this are more normal deposits of shales and limestones, with large Arietid ammonites of th "*bucklandi*" zone.

"The Liassic rocks around Southerndown are difficult to study because the cliffs are not always accessible and the rocks are much disturbed by faults, while fossils, and

* See. T. C. Cantrill in "The Geology of the South Wales Coalfield, Pt. v, The Country around Bridgend," (1904), p. 62.

particularly ammonites, are not by any means abundant. The coral fauna of the Sutton Series is of considerable interest.*" (A. E. Trueman *op. cit.*)

The littoral type also occurs nearer Cardiff in the St. Fagans outlier, in which the rocks consist of massive oolitic limestones and of thick-bedded limestones with many beds of chert-nodules, as seen in quarries one-third of a mile east-south-east of Pentrebanau. The occurrence of this littoral type so close to the normal deposits of Lavernock and Penarth shows that on going northwards towards the coalfield margin littoral conditions develop more quickly than in going westwards.

"The lateral passage exhibited by the Liassic rocks of Glamorganshire is in some respects comparable with that seen in Somersetshire, the lower beds of Liassic in that district where they rest on the Carboniferous Limestone, as for instance around Shepton Mallet, being similar in appearance to the Sutton Stone. So that as the Lower Lias is traced southwards from Gloucestershire to the Mendips, there is a relative increase in the proportion of limestone to shale. Nevertheless there are certain differences in the lateral changes seen in Glamorgan and Somerset respectively, for while the Palæozoic rocks of South Wales were at least relatively stable during the deposition of the Lower Lias, the Mendip regions was subject to considerable movement, and the zones on the Lower Lias in that district are extremely attenuated, partly, perhaps, on account of paucity of sediment, but especially as a result of contemporaneous erosion. Thus the zones, which are not less than 300 feet thick in South Wales, are represented at Radstock by only about 10 feet of sediment, while even so far north as Keynsham, some 10 miles from the Mendip axis, they are not more than about 50 feet thick.

"Although these zones undoubtedly increase considerably in thickness still further to the north, it is doubtful whether

* P. M. Duncan, "On the Astrocœniæ of the Sutton Stone and other Deposits of the Infra-Lias of South Wales," *Quart. Journ. Geol. Soc.*, vol xlii. (1886), p. 101, and various other papers.

anywhere in the Midlands they attain so great a thickness as in South Wales. Similarly, when the corresponding zones are traced southwards from the Mendips, they show a great expansion, but even so far South as Lyme Regis they are little more than half as thick as the rocks of the same age in Glamorgan." (A. E. Trueman *op. cit.*)

Only the lower half of the Lower Lias is seen in South Wales, and no higher Mesozoic Beds or Tertiary Beds are preserved. If such were ever present they have been removed by subsequent denudation.

6. Quaternary Deposits: Glacial and Recent.

A very brief reference to the Glacial and Recent deposits will suffice. The grooving and polishing effects of ice action are visible on some of the exposed rock-surfaces on the Carboniferous Limestone and Millstone Grit outcrops along the northern margin of the coalfield. The direction of ice movement was evidently down the dip slopes from the Brecon Beacons. Boulder Clay covers much of the lower ground in the north of the Coalfield and occurs in patches in the central valleys. Large tracts in the Vale of the Glamorgan are also drift covered. The drift has now been largely scoured out from the deeper valleys of the coalfield by post-glacial river action, and the materials have been spread out as extensive gravel flats over much of the low land south of the Coalfield.

The gravels show a terraced arrangement, one thick sheet extends from Radyr to Llandaff. A large part of Cardiff, including Cathays Park, stands upon a lower terrace 30 to 50 feet above O.D., the gravel varying from 8 to 20 feet thick. The terrace rises gently northwards.

Along the estuary from Penarth eastwards are great level stretches of estuarine mud and alluvium, most of which is of post-Neolithic date. Excavations in the Cardiff and Barry Docks have shown the modern silts to rest upon older silts with intercalated peat beds which extend down to 35 feet below present Ordnance Datum. The lowest peat contains

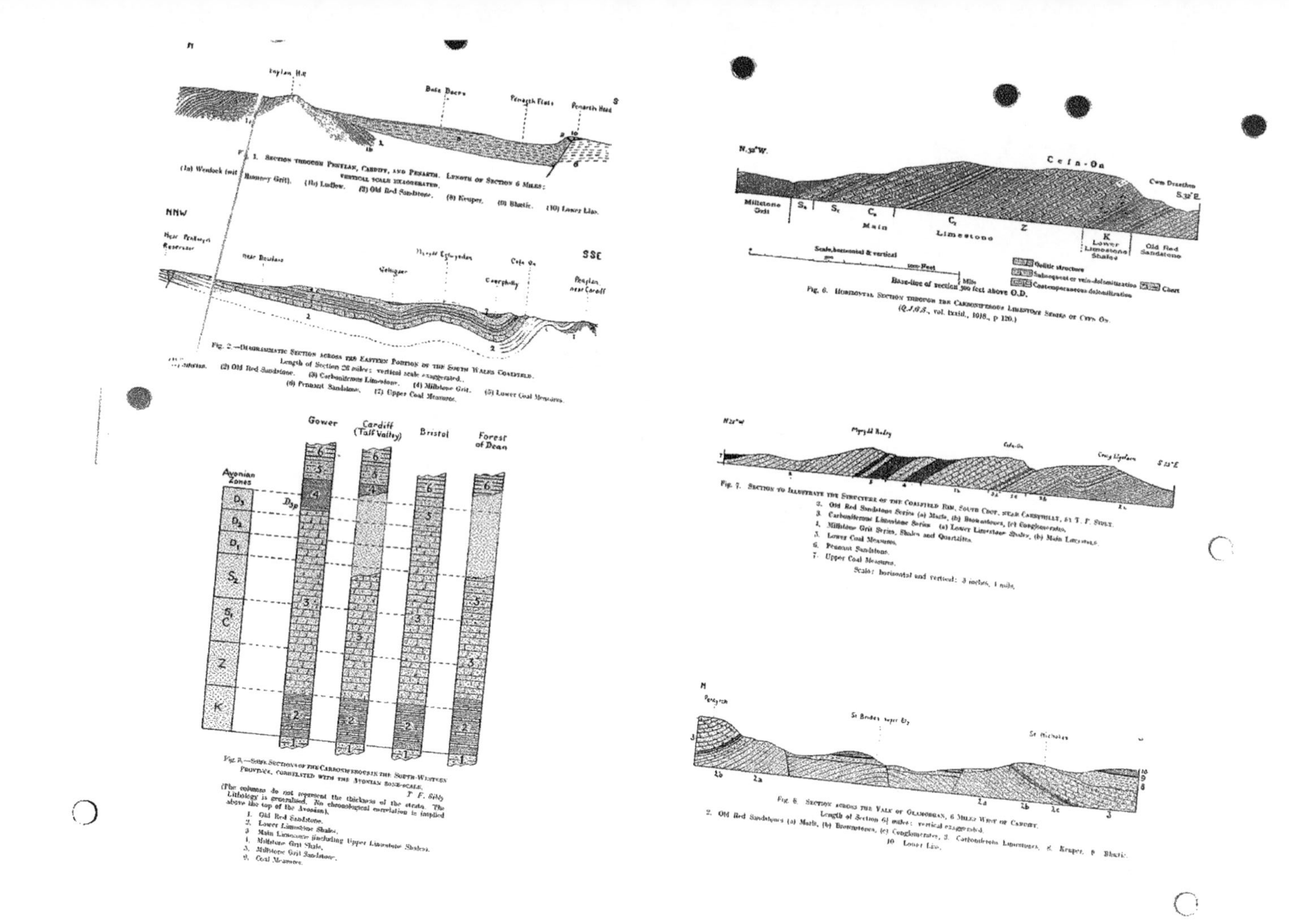

Fig. 1. Section through Penylan, Cardiff, and Penarth. Length of Section 6 miles: vertical scale exaggerated. (1a) Wenlock (with Rumney Grit). (1b) Ludlow. (2) Old Red Sandstone. (8) Keuper. (9) Rhætic. (10) Lower Lias.

Fig. 2.—Diagrammatic Section across the Eastern Portion of the South Wales Coalfield. Length of Section 26 miles: vertical scale exaggerated. (1) Silurian. (2) Old Red Sandstone. (3) Carboniferous Limestone. (4) Millstone Grit. (5) Lower Coal Measures. (6) Pennant Sandstone. (7) Upper Coal Measures.

Fig. 3.—Some Sections of the Carboniferous in the South-Western Province, correlated with the Avonian zone-scale. T. F. Sibly.
(The columns do not represent the thickness of the strata. The Lithology is generalised. No chronological correlation is implied above the top of the Avonian.)
1. Old Red Sandstone.
2. Lower Limestone Shales.
3. Main Limestone (including Upper Limestone Shales).
4. Millstone Grit Shale.
5. Millstone Grit Sandstone.
6. Coal Measures.

Fig. 6. Horizontal Section through the Carboniferous Limestone Series of Cefn On. (Q.J.G.S., vol. lxxiii., 1918, p 120.)

Fig. 7. Section to illustrate the Structure of the Coalfield Rim, South Crop, near Caerphilly, by T. F. Sibly.
2. Old Red Sandstone Series (a) Marls, (b) Brownstones, (c) Conglomerates.
3. Carboniferous Limestone Series (a) Lower Limestone Shales, (b) Main Limestone.
4. Millstone Grit Series, Shales and Quartzites.
5. Lower Coal Measures.
6. Pennant Sandstone.
7. Upper Coal Measures.
Scale: horizontal and vertical: 3 inches, 1 mile.

Fig. 8. Section across the Vale of Glamorgan, 6 Miles West of Cardiff. Length of Section 6½ miles: vertical exaggerated.
2. Old Red Sandstone (a) Marls, (b) Brownstones, (c) Conglomerates, 3. Carboniferous Limestones, 8. Keuper, 9. Rhætic, 10. Lower Lias.

various tree stumps rooted in an underlying soil with land shells, so that a subsidence of not less than 55 feet must have taken place since the peat bed was formed. The presence of polished flint implements and bone needles in higher peats proves that the subsidence was in progress in Neolithic times. Antlers of red-deer and bones of various prehistoric animals, together with an occasional human bone, have been obtained from the lower alluvial beds. (See *Cardiff Memoir*, p. 81, *et. seq.*)

LIST OF BOOKS

ON THE NATURAL HISTORY, TOPOGRAPHY AND ANTIQUITIES OF CARDIFF AND DISTRICT.

NOTE.—This selected list of books, pamphlets, and contributions to periodicals, which are likely to be of interest to scientific visitors to Cardiff, has been compiled by Mr. Wyndham Morgan, F.L.A., the Chief Cataloguer of the Cardiff Public Library. All the books and papers in the list may be consulted in the Reference Library.

I. NATURAL HISTORY.

A. *Meteorology.*

CAPPER (James). *Manuscript.* Meteorological Journal kept [by Colonel Capper] at Cathay near Cardiff, Glamorganshire. Commencing May 1st, 1800. [To May 16th, 1807.] MS. 2.298

— General observations on the height of the thermometer at Cathay [Cardiff]. [*In* Meteorological and miscellaneous tracts. By J. Capper. [1812 ?] Pp. 164-73.] W. 4.1074

EVANS (F. G.) Monthly meteorological notes. 1870-96. [*In* Trans. Cardiff Naturalists' Soc. Vols. i.-xxix.] D. 2.44

HEYWOOD (H.) Meteorological observations in the Society's district. 1897-1908. [*In* Trans. Cardiff Naturalists' Soc. Vols. xxx.-xlii.] D. 2.44

WALFORD (E.) Meteorological observations in the Society's district. 1909-1916. [*In* Trans. Cardiff Naturalists' Soc. Vols. xliii.-xlix.] D. 2.44

B. *Geology.*

ADAMS (William). On the geological features of the South Wales Coalfield. 1870. 12 pp. Table. Sm. 8° W 4.917

— On the Penarth or rhætic beds of Glamorganshire, 1876. [*In* Trans. Cardiff Naturalists' Soc. Vol. viii., pp. 96-98] D 2.44

CANTRILL (T. C.) Geological notes on the excavations at the Gatehouse, Llantwit Major, July and August, 1913. 1914. [*In* Trans. Cardiff Naturalists' Soc. Vol. xlvii., pp. 42-4.] D 2.44

DIXEY (F.) & SIBLY (T. F.) The carboniferous limestone series on the South-Eastern margin of the South Wales coalfield. 1917. [*In* Journal. Geological Society of London. Vol. 73, pp. 111-164.] D 2.463

ETHERIDGE (R.) On the physical structure and organic remains of the Penarth (rhætic) beds of Penarth and Lavernock. 1870 [*In* Trans. Cardiff Naturalists' Soc. Vol. iii. Pt. ii., pp. 39-64.] D 2.44

GEOLOGICAL SURVEY OF ENGLAND AND WALES. Memoirs. 1899-1917. 10 vols. Diagrams, tables. 8° W 5.158

Geology of the South Wales Coalfield—

Pt. 1. Newport district. 1899.

—— 2nd ed. 1909.

Pt. 2. Abergavenny District. 1900.

Pt. 3. Cardiff district. 1902.

—— 2nd ed. 1912.

Pt. 4. Pontypridd and Maesteg district. 1903.

—— 2nd ed. 1917.

Pt. 5. Merthyr Tydfil district. 1904.

Pt. 6. Bridgend district. 1904.

Maps. Cardiff District. 1in., 6in.

HOWARD (F. T.) Geological notes on the old red sandstone country of Monmouth and Brecon, 1902. 3 pp. Sm. 8° W 4.3362

— The geology of the last Barry Dock with notes on the supposed coalfields of South Glamorgan. 1895. [*In* Trans. Cardiff Naturalists' Soc. Vol. xxviii., pp. 77-89] D 2.44

— The geology of Barry Island. Cardiff, 1895. 14 pp. 2 pl., diags. Sm. 8° W 4.434

— The geology of the Cowbridge district. 1897. [*In* Trans. Cardiff Naturalists' Soc. Vol. xxx., pp. 36-47]. D 2.44

— Notes on the base of the rhaetic series at Lavernock Point. Cardiff, 1897. 3 pp. Sm. 8° W 4.1506

HOWARD (F. T.) The origin of the physical features of South Wales: and notes on glacial action in Brecknockshire and adjoining districts. Cardiff, [1904]. 49 pp. Maps, diags. 8° W 4.3643

JORDAN (Henry K.) Notes on the South trough of the coal-field, East Glamorgan. 1903. 26 pp. Diags. 8° W 5.639

MOORE (Charles). The liassic and other secondary deposits of the Southerndown Series. 1876. [*In* Trans. Cardiff Naturalists' Soc. Vol. viii., pp. 53-60.] D 2.44

NORTH (F. T.) On a boring for water at Roath, Cardiff; with a note on the underground structure of the pre-triassic rocks of the vicinity 1915. [*In* Trans. Cardiff Naturalists' Soc. Vol. xlviii., pp. 36-49.] D 2.44

SOLLAS (W. J.) On striated pebbles from the triassic conglomerate near Portskewet, Monmouth. 1881. [*In* Geological Magazine. New ser. Decade ii. Vol. 8., pp. 79-80.] D 2.462

— On the Silurian district of Rhymney and Pen-y-lan, Cardiff. 1879. [*In* Trans. Cardiff Naturalists' Soc. Vol. xi., pp. 7-41.] D 2.44

STORRIE (John). Geology of the Roath Park, Cardiff. 1907. [*In* Trans. Cardiff Naturalists' Soc. Vol. xl., pp. 18-25.] D 2.44

STRAHAN (Aubrey). The geology of South Wales. 52 pp. Maps. 8° W 3.4769

Reprinted from "Geology in the field," Jubilee vol., Geologists' Association.

— On the origin of the river-system of South Wales. 1902. [*In* Journal, Geological Society of London. Vol. 58, pp. 207-225.] D 2.463

C. *Palaeontology.*

HOWARD (F. T.) Geological notes. Ichthyosaurus jaw from Lower Penarth, and a deep well boring at Cardiff Docks. Cardiff, 1896. 4 pp. Sm. 8° W 4.373

JONES (T. Rupert). On some bivalved entomostraca from the coal-measures of South Wales. 1868. [*In* Trans. Cardiff Naturalists' Soc. Vol. ii., pp. 112-118] D 2.44

OWEN (*Sir* Richard). Description of some remains of an air-breathing vertebrate (*Anthrakerpeton crassosteum*, *ow*). From the coal-shale of Glamorganshire. 1868. [*In* Trans. Cardiff Naturalists' Soc. Vol. ii., pp. 108-111.] D 2.44

SOLLAS (W. J.) On some three-toed footprints from the triassic conglomerate of South Wales. 1879. [*In* Jour. Geological Soc. of London. Vol. 31., pp. 511-6.] D 2.463

THOMAS (Thomas Henry). Tridactyl uniserial ichnolites in the trias at Newton Nottage, near Porthcawl, Glamorganshire. 1879. [*In* Trans. Cardiff Naturalists' Soc. Vol. x., pp. 72-91.] D 2.44

WOODWARD (Harry). On a collection of trilobites . . . and one from Glamorganshire. 1902. [*In* Geological Magazine. New ser. Decade iv. Vol. ix., pp. 481-7.] D 2.462

WOTTON (F. W.) Land and fresh water shells of Cardiff. 1888. [*In* Trans. Cardiff Naturalists' Soc. Vol. xx., pp. 31-39.] D 2.44

D. Mineralogy.

BOOKER (Thomas William). Gwent and Dyfed Royal Eisteddfod, 1834. The prize treatise on the mineral basin of Glamorgan and the adjoining district, and the national benefits arising therefrom. Cardiff, 1834. 51 pp. Sm. 8° W 4.921

DAVIES (Henry). The South Wales coalfield; its geology and mines. Pontypridd, 1901. 148 pp. Illus. 8° W 4.2778

DAVIES (L. J.) South Wales coals: their analyses, chemistry and geology. Cardiff, 1920. 99 pp. 8° C 1212

E. Flora.

BLOSSE (*Sir* Robert Lynch). Catalogus Plantarum, In Respositorio Gabalvæ. Cardiff, 1808. 18 pp. Obl. 4° W 6.41

CARDIFF. PARKS COMMITTEE. Guide to Roath Park and catalogue of herbaceous plants in Botanical Gardens; [by W. W. Pettigrew.] Cardiff, [1905]. 67 pp. Illus., map. Sm. 8° W 3.6114

DILLWYN (L. W.) Materials for a fauna and flora of Swansea and the neighbourhood. Swansea, 1848. 44 pp. 4° W 5.42

FAULKNER (R. W.) Contributions towards a catalogue of plants indigenous to the neighbourhood of Tenby. London. 1848. 54 pp. 8° W 3.862

HAMILTON (S.) The flora of Monmouthshire: being a catalogue of all the flowering plants known to be native to the County; and a guide to localities, together with a list of ferns. Newport, 1909. 81, xi. pp. Sm. 8° W. 3.7208

HARRIS (Henry). The flora of the Rhondda; compiled for the Rhondda Naturalists' Society. Bristol, 1905. 86 pp. 8° W 3.5596

RIDDELSDELL (H. J.) A flora of Glamorganshire. (*Reprinted from the "Journal of Botany,"* 1907.) London, 1907. 88 pp. 8° W 4.5353

— Lightfoot's journey to Wales in 1773. 1915. [*In* Jour. of Botany. Vol. 43, pp. 290-307.] D 2.460

SHOOLBRED (W. A.) The flora of Chepstow. 1920. x., 140 pp. Map. 8° W 4.6409

STORRIE (John). The flora of Cardiff: a descriptive list of the indigenous plants found in the district of the Cardiff Naturalists' Society, with a list of the other British and exotic species, found on Cardiff ballast hills. Cardiff, 1886. v., 129 pp. 8° W 4.805

— Notes on the ballast plants of Cardiff and neighbourhood, 1876. [*In* Trans. Cardiff Naturalists' Soc. Vol. viii., pp. 141-145.] D 2.44

— Notes on the flora of the Steep Holme. 1877. [*In* Trans. Cardiff Naturalists' Soc. Vol. ix., pp. 53-54.] D 2.44

THOMAS (Thomas Henry). Notes upon some fine specimens of oak, yew, elm and beech, chiefly in the counties of Monmouth and Glamorgan. 1880. [*In* Trans. Cardiff Naturalists' Soc. Vol. xi., pp. 15-24.] D 2.44

TROW (A. H.), *ed.* The flora of Glamorgan. 1911. [*In* Trans. Cardiff Naturalists' Soc. Vol. l., pp. 1-209.] D 2.44

— The flora of Glamorgan, including the spermaphytes and vascular cryptogams, with index. (*Prepared under the direction of a Committee of the Cardiff Naturalists' Society.*) Vol. i. Cardiff, 1911. 209 pp. 8° W 4.5332

TURNER (Godfrey) & Dillwyn (L. W.) English and Welsh botanical guide. London, 1805. 2v. 8° W 3.4699

VACHELL (Charles Tanfield). Contribution towards an account of the narcissi of South Wales. Cardiff, 1894. 16 pp. 8° W 4.381

F. Fauna.

CARDIFF NATURALISTS' SOCIETY. The birds of Glamorgan. 1900. 164 pp. Illus. 4° W 6.483

— The birds of Glamorgan. 1898. [*In* Trans. Cardiff Naturalists' Soc. Vol. xxi., pp. 1-37.] D 2.44

DEAN (J. Davy). Land Mollusca in the Vale of Glamorgan. 1915. [*In* Trans. Cardiff Naturalists' Soc. Vol. xlviii., pp. 50-58] D 2.44

DILLWYN (L. W.) Materials for a fauna and flora of Swansea and neighbourhood. Swansea. 1848. 44 pp. 4° W 5.42

HALLETT (H. M.) Entomological notes. 1914-16 [*In* Trans. Cardiff Naturalists' Soc. Vols. xlvii.-xlix.] D 2.44

— The Hemiptera of Glamorgan. 1916. [*In* Trans. Cardiff Naturalists' Soc. Vol. xlix., pp. 52-63.] D 2.44

— The Hymenoptera Aculeata recorded for the County of Glamorgan. 1911. [*In* Trans. Cardiff Naturalists' Soc. Vol. xlix., pp. 92-9.] D 2.44

JACKSON (A. R.) A contribution to the spider fauna of the County of Glamorgan. 1906. [*In* Trans. Cardiff Naturalists' Soc. Vol. xxxix., pp. 63-77.] D 2.44

MITCHELL (Archibald). The pinewood wasp (*Sirex Inbencus*) and its occurrence at Dunraven. 1897. [*In* Trans. Cardiff Naturalists' Soc. Vol. xxix., pp. 69-73]. D 2.44

MULLENS (W. H.) *and others.* A geographical bibliography of British ornithology. Parts 3-5. London, 1920. 8°
Glamorganshire, pp. 393-6.
Monmouthshire, pp. 197-8.
D 2.826

NEALE (J. J.) Pisces. 1891. [*In* British Association Handbook. Cardiff, 1891. Pp. 193-199.] W 3.1219

— Surface fishes of the Bristol Channel. 1890. [*In* Trans. Cardiff Naturalists' Soc. Vol. xix., pt. 1, pp. 8-19.] D 2.44

— Trawl fishes of the Bristol Channel. 1887. [*In* Trans. Cardiff Naturalists' Soc. Vol. xix., pt. 1, pp. 111-120.] D 2.44

NICHOLL (Digby S. W.) Notes on the rarer birds of Glamorganshire. London, 1889. 23 pp. Sm. 8°. W 4.619

PROGER (T. W.) The wild mammals of Glamorgan. 1912. [*In* Trans. Cardiff Naturalists' Soc. Vol. xlv., pp. 59-69.] D 2.44

PROGER (T. W.) *and others.* Ornithological notes. 1899-1916. [*In* Trans. Cardiff Naturalists' Soc. Vols. xxxii.-xlix.] D2.44

TOMLIN (J. R. le B.) The Coleoptera of Glamorgan. 1912-1915. [*In* Trans. Cardiff Naturalists' Soc. Vols. xlv.-xlviii.] D 2.44

II. HISTORY AND TOPOGRAPHY.

A. Topography.

1. *South Wales.*

BLACK'S Guide to South Wales. 10th ed. 1901. vi., 243 pp. 19 maps. Sm. 8° H 6398

BRADLEY (A. G.) Highways and byways in South Wales. 1903. 420 pp. Illus. 8° W 3.4965

— In the march and borderland of Wales. 1905. 446 pp. 8° W 5.609

RHYS (Ernest). The South Wales coast from Chepstow to Aberystwyth. 1911. 390 pp. Illus. 8° W 3.7752

South Wales: the country of castles. New ed. 1907. 133 pp. 8° W 4.4255

Wade (G. W. & J. H.) South Wales. 1913. xiii., 295 pp. Illus., maps. Sm. 8° (*Little Guides*) W 2.4930

2. *Glamorgan.*

a. General.

Bradley (A. G.) Glamorgan and Gower. 1908. xii., 146 pp. Illus., map. 8° K 1713

Clark (G. T.) Cartæ et alia munimenta quæ ad Dominium de Glamorgancia pertinent. Dowlais and Cardiff, 1885-93. 4 vols. 4° W 6.80

—— New ed. Cardiff, 1910. 6 vols. 4° W 5.933

— The land of Morgan. London, 1883. 166 pp. 8° W 4.219

— Mediæval military architecture in England. London, 1884. 2 vols. 8° W 5.39

Many of the castles of Glamorgan are dealt with.

Davies (J. D.) A little England beyond Wales; a popular edition of "West Gower," abridged by Dorothy Helme. [1912.] 88 pp. Illus. 8° K 4911

Evans (C. J.) The story of Glamorgan. 1908. xi., 316 pp. Illus., maps. 8° (*Welsh County Series*) H 4065

Farnsworth (C. H.) A school handbook of Glamorgan. Cardiff, 1905. 44 pp. 8° W 3.5405

Isaac (D. L.) Siluriana. Newport, 1859. 336 pp. W 3.1199

Jones (D. W.) *Dafydd Morganwg*. Hanes Morganwg. Aberdar, 1874. 532 pp. 8° W 4.791

Manby (G. W.) Historic and picturesque guide, through the counties of Monmouth, Glamorgan, and Brecknock. Bristol, 1802. 316 pp. 8° W 4.22

Merrick (Rice). A booke of Glamorganshires antiquities, 1578; ed. J. A. Corbett. 2nd ed. London, 1887. 159 pp. 4° W 6.79

Nicholas (T.) The history and antiquities of Glamorganshire and its families. London, 1874. 194 pp. 4° W 5.16

PHILLIPS (J.) An attempt at a concise history of Glamorgan. London, 1879. 102 pp. 4° W 5.28

ROWLANDS (J.) Historical notes of the counties of Glamorgan, etc. Cardiff, 1866. 135 pp. 8° W 3.910

SEWARD (Edwin). Some churches and castles in Glamorgan. 1906. [*In* Trans. Cardiff Naturalists' Soc. Vol. xxxix., pp. 78-95.] D 2.44

SPENCER (M. R.) · Annals of South Glamorgan: historical, legendary and descriptive chapters on some leading places of interest. Carmarthen, [1914]. viii., 269 pp. 8° W 4.5786

WADE (J. H.) Glamorganshire. 1914. xi., 196 pp. Sm. 8° W 2.4789

b. Special.

Caerphilly.

[GLYDE (Thomas)]. A guide to Caerphilly Castle. Cardiff, 1895. 32 pp. 8° W 3.1734

GUIDE to Caerphilly Castle. Cardiff, 1880. 35 pp. 8° W 3.1079

Guide to Caerphilly Castle and the neighbourhood. Cardiff, [1910]. 32 pp. 8° W 3.7513

LLOYD (H.) History of Caerphilly from the earliest period to the present time. Pontypridd, 1900. 86 pp. 8° W 3.3555

MORGAN (Owen) *Morien.* History of Caerphilly Castle: its assailants and defenders. [Pontypridd, 1906.] 158 pp. 8° W 3.5968

Castell Coch.

DRANE (R.) Castell Coch: a gossiping companion to the ruin and its neighbourhood. Cardiff, [1858]. 28 pp. 8° W 3.2223

Dinas Powis.

CORBETT (J. S.) Dinaspowys. 1909. [*In* Trans. Cardiff Naturalists' Soc. Vol. xlii., pp. 70-82.] W 4.577

DAVIES (David). Extract from Iolo's note books, St. Andrews and Dynas Powis in A.D. 1789. 1909. [*In* Trans. Cardiff Naturalists' Soc. Vol. xlii., pp. 83-84.] W 4.577

Ewenny.

FREEMAN (E. A.) Ewenny Priory Church. 1890. 24 pp. 8° W 3.928

TURBERVILL (J. P.) Ewenny Priory, Monastery and Fortress. London, 1902. 20 pp. 8° W 4.2982

Flat Holme.

WOTTON (F. W.) A short historical account of the Flat Holme and its natural history. 1890. [*In* Trans. Cardiff Naturalists' Soc. Vol. xxii., pp. 105-111.] D 2.44

Glyn Neath.

HANDBOOK of the Vale of Neath, its railway and waterfalls [with a map]. Neath, 1852. 79 pp. Sm. 8° W 3.789

YOUNG (W.) Guide to the beauties of Glyn Neath. London, 1835. 85 pp. 8° W 4.104

Kenfig.

GRAY (Thomas). The buried city of Kenfig. London, 1909. 348 pp. 8° W 4.4453

Llandaff.

BIRCH (Walter de Gray). Memorials of the See and Cathedral of Llandaff. Neath, 1912. 427 pp. 4° W 5.1089

COMPTON-DAVIES (W. R.) Historical and pictorial glimpses of Llandaff Cathedral. Cardiff, 1896. 77 pp. 8° W 4.1024

— Peeps at Llandaff Cathedral. 12 pp. 8° W 4.2973

CONYBEARE (W. D.) Llandaff Cathedral. 1849. 22 pp. 8° W 4.1579

FISHBOURNE (E. A.) The ancient Cathedral Close at Llandaff. 1883. [*In* Trans. Cardiff Naturalists' Soc. Vol. xv., pp. 9-12.] D 2.44

FREEMAN (E. A.) Remarks on the architecture of Llandaff Cathedral. London, 1850. 101 pp. 8° W 4.4

JAMES (John H.) A history and survey of the Cathedral Church of Llandaff. Cardiff, 1898. xv., 82 pp. F°. Illus. W 6.393

KEMPSON (F. R.) Notes on the history of Llandaff, with reference to the subject for the Pastoral Staff. Worcester, 1889. 8 pp. 8° W 3.816

WALDRON (Clement). Llandaff Cathedral precincts. 1895-6. [*In* Trans. Cardiff Naturalists' Soc. Vol. xxviii., pp. 16-23.] D 2.44

WILMOTT (E. C. Morgan). The Cathedral Church of Llandaff; a description of the building and a short history of the See. London, 1907. 91 pp. 8° E 709

Llangynwyd.

EVANS (T. C.) *Cadrawd.* History of Llangynwyd Parish. Llanelly, 1887. 192 pp. 8° W 4.647

Llantrisant.

MORGAN (Taliesin). History of Llantrisant, Glamorganshire. Cardiff, 1898. 148 pp. 8° W 3.2224

PHILLIPS (Sem.) The history of the Borough of Llantrisant (Glamorganshire) and its neighbourhood. Bristol, 1866. 112 pp. 8° W 2.247

Llantwit Major.

CORBETT (J. S.) Some notes as to Llantwit Major. 1906. [*In* Trans. Cardiff Naturalists' Soc. Vol. xxxix., pp. 49-62.] D 2.44

FRYER (A. C.) Llantwit Major, a fifth century university. London, 1893. 125 pp. 8° W 4.27

RODGER (John W.) The ecclesiastical buildings of Llantwit Major. 1906. [*In* Trans. Cardiff Naturalists' Soc. Vol. xxxix., pp. 18-48.] D 2.44

TREVELYAN (Marie) *Mrs. Paslieu.* Llantwit Major: its history and antiquities. Newport, [1910]. 224 pp. 8° W 3.7331

Margam.

BIRCH (Walter de Gray). A history of Margam Abbey . . . London, 1897. 405 pp. 4° W 6.269

Merthyr Mawr.

EVANSON (M.) Antiquities on the sandhills at Merthyr Mawr, Glamorganshire. London, 1908. 8 pp. 8° W 4.4371

EVANSON (M,) Stones in the Parish of Merthyr Mawr. Cardiff, 1909. 22 pp. 8° W 4.4464

Merthyr Tydfil.

WILKINS (Charles). The history of Merthyr Tydfil. Merthyr Tydfil, 1867. 372 pp. 4° W 4.702

Morlais Castle.

CLARK (G. T.) Some account of Morlais Castle. [1858.] 18 pp. 8° [*From* Archaeologia Cambrensis. Vol. v., 3rd ser.] W 4.271

Nantgarw.

TURNER (William). The ceramics of Swansea and Nantgarw: a history of the factories. 1897. xii., 349 pp. Col. illus. 8° W 6.169

WARD (John). Billingsley and Pardoe: two Derby "men of mark" and their connection with South Wales. 1896. 32 pp. Illus. 8° W 3.1353

Neath.

PHILLIPS (T. C.) Historical glimpses of Neath Abbey ruins. Neath, 1906. 19 pp. 8° W 4.5060

Newton Nottage.

KNIGHT (H. Hey). Account of Newton Nottage. Tenby, 1853. 79 pp. 8° W 4.347

Pontypridd.

[MORGAN (Owen)] *Morien.* History of Pontypridd and Rhondda Valleys. Pontypridd, 1903. 381 pp. 8° W 3,4810

Porthkerry.

ALLEN (J. Romilly). Notes on Porthkerry Church. 1876. 4 pp. [*From* Archaeologia Cambrensis. Vol. vii., 4th ser., pp. 45-8.] W 4.520

St. Donats.

CLARK (G. T.) Thirteen views of the Castle of St. Donat's, with a notice of the Stradling family. Shrewsbury, 1871. 33 pp. 4° W 6.167

St. Fagans.

DAVID (W.) The antiquities of S. Fagans. The castle. 1877. [*In* Trans. Cardiff Naturalists' Soc. Vol. ix., pp. 73-82.] D 2.44

—— Saint Fagan and his church. 1877. [*In* Trans. Cardiff Naturalists' Soc. Vol. ix., pp. 24-37.] D 2.44

MORGAN (Thomas). An essay on the antiquities of St. Fagans, with its castle. Cardiff, [1866]. 40 pp. 8° W 3.685

3. *Cardiff.*

ALLEN (S. W.) Reminiscences. Cardiff, 1918. 239 pp. 8° W 4.6252

BALLINGER (J.) Guide to Cardiff: city and port. Cardiff, 1908. 120 pp. 8° W 4.4357

— Cardiff: an illustrated handbook. Cardiff, 1896. vi., 245 pp. 8° W 4.435

BOOK (The) of Cardiff. Cardiff, [1912]. 132 pp. 8° (*City and County Readers.*) Illus. W 3.4694.

BRITISH ASSOCIATION. Cardiff, 1891. Handbook for Cardiff and district; ed. Ivor James. Cardiff, 1891. 244 pp. 8° W 3.1219

CARDIFF. Development Committee. Cardiff: a commercial and industrial centre; ed. D. W. Lloyd. Cardiff, 1919. 233 pp. 4° Illus. W 5.1507

CARDIFF. Records Committee. Cardiff Records: being materials for a history of the County Borough from the earliest times; ed. J. Hobson Matthews. Cardiff, 1898-1911. 6 vols. F° Illus. W 6.406

HOWELLS (John). Reminiscences of Cardiff, 1838-40. [*In* The Red Dragon. Vol. v., pp. 218-32.] W 4.1200

JENKINS (W. L.) A history of the town and castle of Cardiff. Cardiff, 1854. 90 pp. 8° W 4.326

TROUNCE (W. J.) "Cardiff in the Fifties." Cardiff, 1918. 96 pp. 8° W 4.6262

WINSTONE (J.) Reminiscences of Old Cardiff. 1883. [*In* Trans. Cardiff Naturalists' Soc. Vol. xv., pp. 60-75.] D 2.44

— Further recollections of Cardiff. 1883. [*In* Trans. Cardiff Naturalists' Soc. Vol. xv., pp. 76-88.] D 2.44

B. Antiquities.

1. *Glamorgan.*

ALLEN (J. Romilly). The cylindrical pillar at Llantwit Major. 1889. 10 pp. 8° [*From* Archaeologia Cambrensis. Vol. vi., 5th ser., pp. 317-326.] W. 4.526

— A description of some cairns on Barry Island. 1873. [*In* Archaeologia Cambrensis. Vol. iv., 4th ser., pp. 188-191.] 8° W 4.518

— The inscribed and sculptured stones at Llantwit Major. London, 1889. 11 pp. 8° W 4.524

— Notice of a mediæval thurible found at Penmaen, Gower. 1891. 5 pp. 8° [*From* Archaeologia Cambrensis. Vol. viii., 5th ser., pp. 161-5] W 4.527

BREWER (J.N.), Llandaff. The history and antiquities of the cathedral. London, [1817]. 17 pp. 4° [*In* British Cathedrals.] W 4.559

EVANS (Franklen P.) The St. Nicholas cromlechs. 1881. [*In* Trans. Cardiff Naturalists' Soc. Vol. xiii., pp. 41-48.] D 2.44

EVANSON (Morgan). Stones in the parish of Merthyr Mawr, Glamorgan. 1908. [*In* Trans. Cardiff Naturalists' Soc. Vol. xli., pp. 19-38.] D 2.44

JAMES (Henry). The excavations at Gelligaer Camp. 1899. [*In* Trans. Cardiff Naturalists' Soc. Vol. xxxi., pp. 80-84.] D 2.44

MORGAN (T.) *Llyfnwy.* An essay on the antiquities of St. Fagan's. Cardiff, [1866]. 40 pp. 8° W 3.685

RILEY (William). Intrenchments and camps on Mynydd Baidan and Mynydd Margam. 1894-5. [*In* Trans. Cardiff Naturalists' Soc. Vol. xxvii., pp. 71-89.] D 2.44

RODGER (John W.) Excavations at Llantwit Major. 1912. [*In* Trans. Cardiff Naturalists' Soc. Vol. xlv., pp. 87-89.] D 2.44

— Exploration of the "Gaer Fach" at Gelligaer. 1908. [*In* Trans. Cardiff Naturalists' Soc. Vol. xli., pp. 39-40.] D 2.44

RODGER (John W.) Llantwit Major excavations. 1914. [*In* Trans. Cardiff Naturalists' Soc. Vol. xlvii., pp. 35-41.] D 2.44

— The stone cross slabs of S. Wales and Monmouthshire. 1911. [*In* Trans. Cardiff Naturalists' Soc. Vol. xliv., pp. 24-64.] D 2.44

SEWARD (Edwin). The Castell field at Graig Llwyn, Lisvane, Glamorgan. 1907. [*In* Trans. Cardiff Naturalists' Soc. Vol. xl., pp. 26-33.] D 2.44

— Discovery of a Roman villa at Llantwit Major. 1888. [*In* Trans. Cardiff Naturalists' Soc. Vol. xx., pt. 2, pp. 49-61.] D 2.44

STORRIE (John). Ancient remains on Ely race-course. 1894. [*In* Trans. Cardiff Naturalists' Soc. Vol. xxvi., pp. 125-128.] D 2.44

— Notes on excavations made . . . at Barry Island. Cardiff. 1896. 71 pp. 8° W 3.784

— Roman iron-making at Ely Race-course. 1894. [*In* Trans. Cardiff Naturalists' Soc. Vol. xxvi., pp. 129-133.] D 2.44

THOMAS (T. H.) "Calvary" crosses, Glamorgan. 1904. [*In* Trans. Cardiff Naturalists' Soc. Vol. xxxvii., pp. 55-65.] D 2.44

— Some account of the pre-Norman inscribed and decorated monumental stones of Glamorganshire, being explanatory notes upon the series of photographs made by Mr. T. Mansel Franklen. 1892. [*In* Trans. Cardiff Naturalists' Soc. Vol. xxv., pp. 34-46.] D 2.44

WARD (John). Castell Morgraig. The situation, exploration, remains. 1905. [*In* Trans. Cardiff Naturalists' Soc. Vol. xxxviii., pp. 20-58.] D 2.44

— Notes on Roman remains in the Society's district. 1908. [*In* Trans. Cardiff Naturalists' Soc. Vol. xli., pp. 41-49.] D 2.44

— The Roman fort at Gellygaer. 1903. [*In* Trans. Cardiff Naturalists' Soc. Vol. xxxv.] D 2.44

— The Roman fort of Gellygaer—the baths. 1909. [*In* Trans. Cardiff Naturalists' Soc. Vol. xlii., pp. 25-69.] D 2.44

WARD (John). The annexe. 1911. [*In* Trans. Cardiff Naturalists' Soc. Vol. xliv., pp. 65-91.] D 2.44

—— 1913. [*In* Trans. Cardiff Naturalists' Soc. Vol. xlvi., pp. 1-20.] D 2.44

2. *Cardiff.*

CLARK (G. T.) Cardiff Castle. 1890. [*In* Archæologia Cambrensis, 5th ser. Vol. 7, pp. 283-92.] W 4.1196

— Some account of Cardiff Castle. 1862. [*In* Archæologia Cambrensis. 3rd ser. Vol. 8, pp. 249-71.] W 4.1196

CONWAY (J. P.) The Black Friars of Cardiff: Recent excavations and discoveries. 1889. [*In* Archaeologia Cambrensis. 5th ser. Vol. 6, pp. 97-105.] W 4.1196

FOWLER (C. B.) The excavations carried out on the site of the Blackfriars Monastery, Cardiff Castle. 1897. [*In* Trans. Cardiff Naturalists' Soc. Vol xxx., pp. 5-15.] D 2.44

ROBINSON (George E.) Roman Cardiff. 1877. [*In* Trans. Cardiff Naturalists' Soc. Vol. ix., pp. 19-23.] D 2.44

WARD (John). Recent discoveries of Roman work at Cardiff Castle. 1913. [*In* Trans. Cardiff Naturalists' Soc. Vol. xlvi., pp. 85-89) .] D 2.44

C. *Folklore.*

GRIFFITHS (P. Rhys). Welsh weather-proverbs. 1894. [*In* Trans. Cardiff Naturalists' Soc. Vol. xxvi., pp. 73-80.] D 2.44

JONES (Edmund), of the Tranch, Pontypool. *Manuscript.* A relation of Numerous and Extraordinary Apparitions of Spirits, good and bad, in the Principality of Wales. 1780. (Wooding Collection) Ms. 2.249

— A relation of apparitions of Spirits in the Principality of Wales. 2nd ed. [Trevecca], 1780. 144 pp. 8° W 4.656

—— [3rd ed.] Newport, 1813. 104 pp. 8° W 2.363

REDWOOD (Charles). The Vale of Glamorgan: scenes and tales among the Welsh. London, 1839. viii., 309 pp. 8° W 3.186

RHYS (*Sir* J.) Sacred wells in Wales. 1893. [*In* Folklore. Vol. iv., pp. 55-79.]

— Welsh fairy tales: ix., Glamorgan, x., Gwent. 1882. [*In* Y Cymmrodor. Vol. v., pp. 124-43.] W 4.1198

ROBERTS (W.) "Mari Lwyd" and its origin. 1896. [*In* Trans. Cardiff Naturalists' Soc. Vol. xxix., pp. 80-93.] D 2.44

SIKES (Wirt). British goblins: Welsk folk-lore, fairy mythology, legends and traditions. London, 1880. xvi., 412 pp. 8° W 4.441

THOMAS (T. H.) Some folk-lore of South Wales. Cardiff, [1904]. 13 pp. 8° W 4.3663

TREVELYAN (Marie), *Mrs. Paslieu.* Folk-lore and folk-stories of Wales: intro. by E. S. Hartland. London, 1909. xii., 350 pp. 8° W 4.4529

WHERRY (B. A.) Miscellaneous notes from Monmouthshire. 1905. [*In* Folklore. Vol. 16, pp. 63-7.]

— Wizardry on the Welsh border. 1904. [*In* Folklore. Vol. 15, pp. 75-86.]